GCSE

CHEMISTRY
for CCEA

Judith Johnston
Theo Laverty

Hodder Murray
A MEMBER OF THE HODDER HEADLINE GROUP

Acknowledgements

The publishers would like to thank the following individuals, institutions and companies for permission to reproduce photographs in this book. Every effort has been made to trace ownership of copyright. The publishers would be happy to make arrangements with any copyright holder whom it has not been possible to contact:

Bruce Coleman/HPH Photography (170 left); Corbis (172 right)/Andy Hibbert (188 right)/Charles & Josette Lenars (28 church)/Chinch Gryniewicz (165)/Chris Newton (47 left)/Cooperphoto (41 cans)/D Boone (28 diamond)/ Duomo (47 middle and right)/Gunter Marx (28 copper dome)/Herrmann Starke (28 glassware)/Justin Hutchinson (41 dishes)/Macduff Everton (28 field)/Richard Cummins (28 boat)/Robert Essel (188 middle)/Todd Gipstein (172 left)/Tony Aruza (28 salt); GSF Picture Library (93, 94); Hodder & Stoughton (102); Life File Andrew Ward (52 left)/Emma Lee (41 bleach and jeans, 52 right, 73 left, 176 right, 181, 188 left, 199)/ Fraser Ralston (28 bridge)/Graham Burns (52 middle)/Jeremy Hoare (176 left)/Liz Beech (thermometer); RSPCA/Cheryl Ertelt (182); Science Photo Library (2 top right, 108)/Andrew Lambert (60 both, 117 middle)/Biosym Technologies (138)/Charles Winters (113 all, 117 left)/Claude Nuridsany & Marie perennou (54 bottom)/Peter Fowler (2 bottom)/Francois Sauze (73 right)/Hank Morgan (206 right)/Jerry Mason (40 right, 127)/James Holmes (177 right)//Kaj Svensson (168)/Klaus Guldbrandsen (178)/ Lawrence Livermore (40 middle)/Lawrence Migdale (32)/Matt Meadows (54 top)/Martin Chillmaide (90)/Martin Bond (177 middle, 206 left)/NASA (40 left)/Oulette & Theroux (207)/Peter Scoones (76)/Russ lappa (117 right)/Sheila Terry (2 top left, 91)/ Simon Lewis (28 fireman)/Tek Image (177 left)/Vaughn Fleming (170 right).

Orders: please contact Bookpoint Ltd, 130 Milton Park, Abingdon, Oxon OX14 4SB.
Telephone: (44) 01235 827720. Fax: (44) 01235 400454. Lines are open from 9.00–6.00, Monday to Saturday, with a 24 hour message answering service.

You can also order through our website www.hodderheadline.co.uk

British Library Cataloguing in Publication Data
A catalogue record for this title is available from the British Library

ISBN 978 0340 858240

First Published 2003
Impression number   10   9   8   7   6
Year   2011   2010   2009   2008   2007

Cover photo from Science Photo Library

Typeset by Tech-Set Limited, Gateshead, Tyne and Wear

Printed in Singapore for Hodder & Stoughton Educational, a division of Hodder Headline Ltd, 338 Euston Road, London NW1 3BH.

Contents

Preface

The GCSE Science for CCEA series comprises of three books: GCSE Biology for CCEA, GCSE Chemistry for CCEA, and GCSE Physics for CCEA, which together cover all aspects of the material needed for students following the CCEA GCSE specifications in:

- Science: Double Award (Modular)
- Science: Double Award (Non-Modular)
- Science: Biology
- Science: Chemistry
- Science: Physics

GCSE Chemistry for CCEA covers all the material relating to the chemistry component of the CCEA Science Double Award (Modular and Non-modular), together with the additional material required for the CCEA Science: Chemistry specification.

Judith Johnston and Theo Laverty are both experienced chemistry teachers.

Identifying Specification and Tier

The material required for each specification and tier is clearly identified using the following colour code:

Material required for foundation tier students following either the Science Double Award (Modular and Non-Modular) or the Science: Chemistry specifications is identified by a red line running down the left-hand side of the text.

Material required for higher tier students following either the Science Double Award (Modular and Non-Modular) or the Science: Chemistry specifications is identified by text with a red tinted background.

Material required for foundation tier students following the Science: Chemistry specification is identified by a blue line running down the left-hand side of the text.

Material required for higher tier students following the Science: Chemistry specification is identified by text with a blue tinted background.

Scientific Investigation

During your course you will be required to carry out a number of scientific investigations. You will need to provide a written report which focuses on the following three skills:

1 Planning – Here you will need to write about what you intend to do. You will need to think clearly about what you are planning to investigate and what apparatus you will need. You will also need to use your scientific knowledge and understanding to plan a procedure, identifying key factors that will need to be either varied, controlled or considered. In addition you will need to make a prediction about what you think your investigation will demonstrate and to justify your prediction. Finally you will need to outline a strategy for dealing with your results.

2 Obtaining evidence – Here you will need to demonstrate that you can collect and record evidence in an accurate and systematic way. Your teacher will want to be sure that you are working safely and that you have checked and repeated your work where necessary. To gain the highest possible marks you will need to demonstrate that you can carry out the work skilfully, and can obtain and record an appropriate range of reliable evidence.

3 Interpreting and evaluating – In this skill area you need to use diagrams, charts or graphs as the basis for explaining the evidence that you have collected. You will be expected to use numerical methods, such as averaging, where necessary. Your teacher will want to be sure that you can draw a valid conclusion, which is consistent with your evidence, and that draws on your knowledge and understanding. In addition you will need to explain the extent to which your conclusion supports the prediction you made in your plan. Finally you will need to consider the reliability of your evidence and whether your procedure could have been improved. Is there enough evidence to support your conclusion? Are there any strange results, and if so can you explain how they arose?

Chapter 1

Atomic Structure

Learning objectives

By the end of this chapter you should be able to:

➤ Recall the different ideas and models put forward to explain the structure of matter and atoms

➤ Understand that the atom has a nucleus composed of protons and neutrons with electrons orbiting the nucleus in shells

➤ Recall the relative charges and masses of protons, neutrons and electrons

➤ Explain the terms atomic number and mass number

➤ Describe the structure of atoms in terms of protons, neutrons and electrons (limited to the first 20 elements in the Periodic Table)

➤ Understand that the electronic structure of an element shows how the electrons are arranged in shells, for example, sodium 2,8,1

➤ Understand that isotopes are atoms of the same element which have the same number of protons but different numbers of neutrons, for example, ^{35}Cl and ^{37}Cl

Theories of atomic structure

Around two thousand years ago the Greek philosophers Democritus and Leucippus, first put forward the idea that matter was composed of small, invisible particles, called **atoms**. Unfortunately their theory was not taken seriously and instead the simpler ideas of another Greek philosopher Aristotle were considered more acceptable. He believed that matter was composed of the four 'elements', air, earth, fire and water and that these four 'elements' could be used to explain matter and its behaviour.

John Dalton's Theory

It was not until 1808 when John Dalton published his Atomic Theory that the significance of the particulate nature of matter was taken seriously by scientists. In his theory Dalton stated that:

● All elements are made up of small indivisible particles called atoms.

● Atoms cannot be created or destroyed.

● Atoms of different elements have different properties.

● When atoms combine they form molecules or compounds (Dalton called these 'compound atoms').

Dalton's research provided important explanations for the structure of compounds and molecules and it was used to determine the atomic masses of the elements.

Figure 1 John Dalton and J. J. Thomson

The 'Plum Pudding' model of J. J. Thomson

It was not until 1897 when J. J. Thomson discovered the electron that Dalton's atomic theory was modified to provide a new model of atomic structure. Using the **Plum Pudding model** of the atom, Thomson put forward the idea that in atoms there are rings of negative **electrons** embedded in a sphere of positive charge, just like currants embedded in a Christmas pudding. Thomson also explained that the atom was neutral since it contained equal numbers of positive and negative charges.

The work of Ernest Rutherford

Although Thomson had made important progress into our understanding of atomic structure, he incorrectly made the assumption that the mass of an atom was only due to electrons. In 1913 his assumption was shown to be incorrect when **Ernest Rutherford** proved that the atom consisted of electrons revolving around a positively charged **nucleus**. He called the positive particles **protons** and gave them a mass of one atomic mass unit. Rutherford's calculations also showed that virtually all of an atom's mass is contained in the nucleus and that there is a large volume of space between the nucleus and the revolving electrons. Rutherford compared his model of electrons revolving around a positive nucleus to that of the planets revolving around the sun. On a relative scale his model of the atom could be compared to a small pea at the centre of a football pitch, where the small pea is the nucleus with the remainder of the pitch representing the space occupied by the revolving electrons.

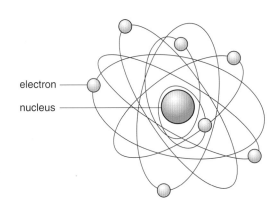

electron
nucleus

Figure 2 Ernest Rutherford and his model of the atom

In 1932 Rutherford's model of the atom was further refined, when **David Chadwick** showed that the nucleus of the atom contained two different types of particles, protons and new particles, called **neutrons**. The neutron was found to have a similar mass to a proton but, unlike the proton, it had no charge. It has been shown that all atoms apart from hydrogen contain neutrons.

Present day ideas of the atom

Present day understanding of atomic structure shows that the atom is composed of a very small, dense, positively charged nucleus composed of protons and neutrons, with negatively charged electrons in shells, orbiting the nucleus. The radius of an atom is very small and only measures about 10^{-8} cm. Table 1 summarises relative mass, charge and position of the three particles that make up the atom.

Particle	Relative mass (atomic mass units)	Relative charge	Position
Neutron	1	0	nucleus
Proton	1	$^+1$	nucleus
Electron	$\dfrac{1}{1840}$	$^-1$	shell

Table 1 Mass, charge and position of protons, electrons and neutrons

From the table it is seen that:

● Protons and neutrons have the same mass and it would take 1840 electrons to have the same mass as either a proton or a neutron.

● Protons and electrons have opposite charges while neutrons have no charge.

● Protons and neutrons are contained in the nucleus while electrons are contained in shells.

Questions

1   Name the scientist who discovered the

    a) electron   b) proton   c) neutron

2   Copy and complete the table below about the particles which make up atoms

Particle	Relative Mass	Charge	Position
?	?	?	shell
?	?	$^+1$	?
neutron	1	?	?

3   **IT:** Use PowerPoint to produce a presentation showing the historical development of the structure of the atom. Your slides should include:

● the work of John Dalton, J J Thompson, Ernest Rutherford and David Chadwick

● the relative charge and mass, and position of the particles which make up the atom

● the present day model of the atom.

Atomic Number and Mass Number

The terms **atomic number** and **mass number** are used to provide scientists with important information about the number of protons, neutrons and electrons contained in atoms.

> The Atomic Number (Z) of an element is the number of protons in an atom of that element.

Since atoms are neutral there are always the same number of electrons and protons in each atom. The atomic number of an element can be obtained from the Periodic Table as is represented by the subscript at the left hand side of the symbol e.g. $_{11}$Na. This tells us that a sodium atom has 11 protons and 11 electrons.

> The Mass Number (A) of an element is the number of protons and neutrons in the nucleus of an atom of that element.

Like the atomic number, the mass number can be obtained from the Periodic Table.

For the element in question it is the superscript at the left hand side of the symbol e.g. ^{23}Na. Using sodium as an example, this means that the number of protons and neutrons added together is 23. The number of neutrons can be obtained by subtracting the atomic number from the mass number i.e. $23 - 11 = 12$ neutrons in an atom of sodium.

Questions

1   Explain the meaning of the terms atomic number and mass number.

2   Use your Periodic Table to work out the mass number and atomic number for

   a) Ca   b) Rb   c) Zn

3   Use your Periodic Table to identify the elements that have the following atomic numbers:

   a) 38   b) 14   c) 2

4   Calculate the number of electrons, protons and neutrons in the following:

   a) $^{19}_{9}$F   b) $^{31}_{15}$P   c) $^{40}_{18}$Ar   d) $^{106}_{46}$Pd   e) $^{56}_{26}$Fe   f) $^{64}_{29}$Cu

5   Why do atoms have no charge?

6   Copy and complete the table below.

Element	Atomic number	Number of neutrons
?	19	20
Manganese	25	?
Bromine	?	45

How electrons are arranged in atoms

It is now understood that the atom is composed of a very small dense, positively charged nucleus composed of protons and neutrons, with negatively charged electrons in shells, orbiting the nucleus. Figure 3 shows how the electrons are arranged in the first shell for the two lightest elements, hydrogen and helium:

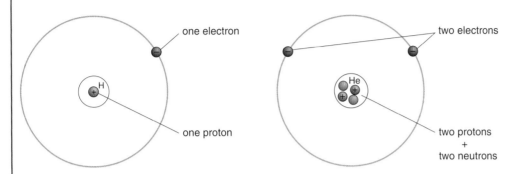

Figure 3 Electronic structures of hydrogen and helium

For GCSE chemistry the following guidelines will apply when filling electrons into shells:

● Electrons are filled into shells starting from the first shell (sometimes called the first energy level), which is the one closest to the nucleus. Moving out from the first shell, electrons are then filled into the second shell, third shell and so forth.
● The first shell can hold up to 2 electrons while other shells can hold up to 8 electrons.
● Before filling electrons into a new shell it is important that the existing shell has been filled.

Filling electrons into shells for nitrogen and sodium

Nitrogen has seven electrons and the first two electrons are filled into the first shell with the remaining five electrons placed in the second shell. For sodium with eleven electrons, two are used to fill the first shell while the next eight electrons fill the second shell and the remaining electron goes into the third shell. These electronic structures are often shortened to nitrogen 2,5 and sodium 2,8,1 as shown in Figure 4.

Nitrogen, $^{14}_{7}$N
7 protons
7 electrons
7 neutrons

Sodium $^{23}_{11}$Na
11 protons
11 electrons
12 neutrons

Nitrogen

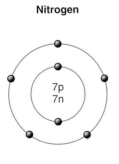

electronic structure: 2, 5

Sodium

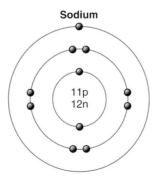

electronic structure: 2, 8, 1

Figure 4 Electronic structures of nitrogen and sodium

Structure of the first twenty elements in the Periodic Table

In Table 2 the Atomic Number and Mass Number of the first 20 elements in the Periodic Table have been used to work out the number of protons, electrons and neutrons which each element contains. The last column provides a description of each element's electronic structure in terms of electrons in shells.

Element	Symbol	Atomic number	Mass number	Number of protons	Number of electrons	Number of neutrons	Electronic structure (electrons in shells)
Hydrogen	$^{1}_{1}H$	1	1	1	1	0	1
Helium	$^{4}_{2}He$	2	4	2	2	2	2
							First shell is full
Lithium	$^{7}_{3}Li$	3	7	3	3	4	2,1
Beryllium	$^{9}_{4}Be$	4	9	4	4	5	2,2
Boron	$^{11}_{5}B$	5	11	5	5	6	2,3
Carbon	$^{12}_{6}C$	6	12	6	6	6	2,4
Nitrogen	$^{14}_{7}N$	7	14	7	7	7	2,5
Oxygen	$^{16}_{8}O$	8	16	8	8	8	2,6
Fluorine	$^{19}_{9}F$	9	19	9	9	10	2,7
Neon	$^{20}_{10}Ne$	10	20	10	10	10	2,8
							Second shell is full
Sodium	$^{23}_{11}Na$	11	23	11	11	12	2,8,1
Magnesium	$^{24}_{12}Mg$	12	24	12	12	12	2,8,2
Aluminium	$^{27}_{13}Al$	13	27	13	13	14	2,8,3
Silicon	$^{28}_{14}Si$	14	28	14	14	14	2,8,4
Phosphorus	$^{31}_{15}P$	15	31	15	15	16	2,8,5
Sulphur	$^{32}_{16}S$	16	32	16	16	16	2,8,6
Chlorine	$^{35}_{17}Cl$	17	35 or 37*	17	17	18 or 20	2,8,7
Argon	$^{40}_{18}Ar$	18	40	18	18	22	2,8,8
							Third shell is full
Potassium	$^{39}_{19}K$	19	39	19	19	20	2,8,8,1
Calcium	$^{40}_{20}Ca$	20	40	20	20	20	2,8,8,2

(*chlorine has two isotopes as shown on page 7)

Table 2 Electronic structures of the first 20 elements

Questions

1 Draw diagrams to show how the electrons, protons and neutrons are arranged in:

a) $^{7}_{3}\text{Li}$ b) $^{32}_{16}\text{S}$ c) $^{9}_{4}\text{Be}$ d) $^{35}_{17}\text{Cl}$

2 From the following table select an element which has:

a) a filled outer shell of electrons

b) four electrons in its outer shell

c) six electrons in its outer shell.

Element	Mass number	Number of neutrons
A	9	5
D	12	6
E	40	22
G	39	20
H	31	15

3 Copy and complete the table below.

Element	Mass Number	Atomic Number	Electronic Structure
?	27	?	2.8.3
?	24	12	?
Fluorine	?	9	?
?	16	?	2.6

Isotopes

Atoms of a particular element always have the same atomic number which means that they have the same number of protons and electrons; however, for many elements their atoms can have different masses. When this occurs elements are said to have **isotopes**.

Isotopes are atoms of the same element which have the same number of protons but different numbers of neutrons.

Isotope	Protons	Electrons	Neutrons
$^{35}_{17}\text{Cl}$	17	17	18
$^{37}_{17}\text{Cl}$	17	17	20
$^{1}_{1}\text{H}$	1	1	0
$^{2}_{1}\text{H}$	1	1	1
$^{3}_{1}\text{H}$	1	1	2

Table 3 Isotopes of chlorine and hydrogen

From the definition it is seen that isotopes have the same atomic number but different mass number. Table 3 shows the common isotopes of two elements, chlorine and hydrogen. Taking chlorine as an example it is seen that chlorine is made up of two isotopes; chlorine-35 and chlorine-37. Because both atoms contain the same number of electrons, arranged in the same way, it means that their chemical reactivity is identical. The only difference between the isotopes is that chlorine-35 is a lighter atom with two fewer neutrons. Consequently, the isotopes have different physical properties; for example, Cl-37 has a greater density than Cl-35.

It has been shown that naturally occurring chlorine is made up of 75% chlorine-35 and 25% chlorine-37 and from the following calculation it is seen that the average relative mass for a chlorine atom is 35.5:

$$\text{Average relative mass} = \frac{35 \times 75 + 37 \times 25}{100} = 35.5$$

The average relative mass which accounts for the different masses of the isotopes and also their relative amounts in a naturally occurring sample is called **the relative atomic mass**.

Atomic structures of the isotopes of hydrogen

Figure 5 shows the atomic structures of the three isotopes of hydrogen. While the electronic structure of each isotope is the same, it is seen that the nuclei of the atoms are different due to the different number of neutrons for each isotopes.

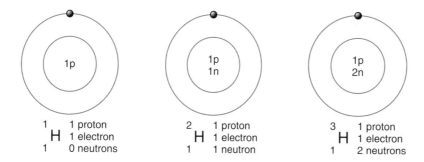

$^{1}_{1}\text{H}$	1 proton / 1 electron / 0 neutrons
$^{2}_{1}\text{H}$	1 proton / 1 electron / 1 neutron
$^{3}_{1}\text{H}$	1 proton / 1 electron / 2 neutrons

Figure 5 Atomic structures of the isotopes of hydrogen

Questions

1 Explain the meaning of the term *isotopes*.

2 Copy and complete the table to show the number of protons, electrons and neutrons in the three isotopes of carbon.

Isotope	Electrons	Protons	Neutrons
$^{12}_{6}\text{C}$	6	?	?
$^{13}_{6}\text{C}$	?	6	?
$^{14}_{6}\text{C}$	?	?	8

3 Chlorine exists naturally as two isotopes, $^{35}_{17}\text{Cl}$ and $^{37}_{17}\text{Cl}$ in the ratio 3 : 1 respectively.

Show why the relative atomic mass of chlorine is 35.5.

4 Calculate the relative atomic mass of boron given that it contains 20% boron-10 and 80% boron-11.

5 **IT**: Use Publisher to design a poster on isotopes. Your poster should include diagrams to show how the particles are arranged in atoms of isotopes. Write a few sentences to explain the meaning of your diagrams.

Websites

www.watertown.k12.wi.us/hs/teachers/buescher/atomtime.asp

WWW

Exam questions

1 a) Copy the diagram and label it to show where the protons, neutrons and electrons are in this atom.

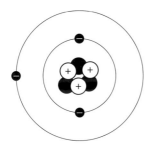

(3 marks)

b) Name the element shown above. You may find your Data Leaflet useful.

(1 mark)

2 The table below gives details about some elements in Group II of the Periodic Table.

Use your Data Leaflet to copy and complete this table.

Symbol	Relative atomic mass	Atomic number
Be	9	
Mg	24	12
Ca	40	20
	88	38
Ba		56

(3 marks)

3 Use the Periodic Table in your **Data Leaflet** to help you fill in the blanks in the table below.

Element	Relative atomic mass	Number of protons	Number of electrons	Number of neutrons
Boron	11	5		6
	39		19	20
Lithium	7	3	3	

(3 marks)

4 This question is about the elements in Group II of the Periodic Table.

You may use the Periodic Table in your Data Leaflet to answer this question.

a) How many **electrons** does a calcium atom have?

(1 mark)

b) How many **protons** does a magnesium atom have?

(1 mark)

c) Which Group II element can have five **neutrons**?

(1 mark)

5 The following table contains information about the structure of some particles which are either atoms or ions.

Particle	Number of protons	Number of neutrons	Number of electrons
A	17	18	17
B	16	16	16
C	19	20	18
D	17	20	17
E	19	20	19

a) Which **one** of the following is **not** found in the nucleus of an atom?

    proton   **neutron**   **electron**

(1 mark)

b) What is the mass number of particle A?

(1 mark)

c) Give the letters of the **two** particles which are **isotopes**.

(1 mark)

6 a) Copy and complete the table below which gives information about protons, neutrons and electrons.

Particle	Relative charge	Relative mass
Proton	+1	
Neutron		1
Electron		$\dfrac{1}{1840}$

(3 marks)

b) Uranium has **two** common isotopes

$$^{235}_{92}U \text{ and } ^{238}_{92}U$$

(i) What is the atomic number of uranium?

(1 mark)

(ii) Why does the $^{238}_{92}U$ isotope have a greater mass number than the $^{235}_{92}U$ isotope?

(1 mark)

(iii) Give **one** way in which atoms of the two uranium isotopes are similar.

(1 mark)

7 Chlorine exists as **two** isotopes $^{35}_{17}Cl$ and $^{37}_{17}Cl$.

a) (i) Give **one** way in which these isotopes are **similar**.

(1 mark)

(ii) Give **one** way in which these two isotopes are **different**.

(1 mark)

Hydrogen exists as **three** isotopes. One of its isotopes is $^{1}_{1}H$.

b) Which of the following are the two correct isotopes of hydrogen.

    $^{2}_{1}H$ $^{1}_{2}H$ $^{3}_{1}H$ $^{2}_{0}H$

(2 marks)

Chapter 2

Chemical Bonding and the Properties of Materials

Learning objectives

By the end of this chapter you should be able to:

➤ Recall that a compound is a substance which contains two or more elements chemically joined together

➤ Describe the structure of ions in terms of protons, neutrons and electrons (limited to the first twenty elements)

➤ Describe the formation of an ionic bond in terms of electron transfer and recognise that the ionic bond is the attraction between ions of opposite charge

➤ Recall that ionic bonding is typical of metal compounds: for example, MgO, NaCl and $CaCl_2$

➤ Describe the formation of a covalent bond in terms of sharing electron pairs for example Cl_2, O_2, H_2O and CH_4

➤ Recognise covalent bonding as typical of non-metal elements and compounds

➤ Understand that a molecule is a group of atoms held together by covalent bonding

➤ Understand that covalent substances can have a molecular or a giant molecular structure

➤ Describe the bonding in metals

➤ Classify substances in terms of their properties as ionic; metallic; covalent molecular or giant molecular (to include graphite, diamond and quartz)

➤ Explain the properties and uses of typical ionic, covalent (simple and giant) and metallic substances in terms of their chemical bonding and structures

The three types of bonding

To understand the properties and uses of materials it is important to know how elements combine together and how the atoms are arranged in the new substances. To do this it is necessary to study the different types of chemical bonding that can exist between the atoms of elements. The three main types of bonding are:

● **Ionic** bonding which occurs between metals and non-metals; for example, when magnesium metal reacts with the non-metal oxygen to produce magnesium oxide.

● **Covalent** bonding that takes place between non-metals, for example, hydrogen combining with oxygen to produce water.

● **Metallic** bonding which is formed between the atoms of a metal, for example, copper.

At this stage it is important to recap on the meaning of the three terms atom, element and compound as they will be used throughout the discussion on bonding.

- An **atom** is the smallest part of an element which can exist, for example copper metal is composed of copper atoms.
- An **element** is a substance composed of one type of atom only, for example copper or oxygen or bromine.
- A **compound** is a substance which contains two or more elements chemically joined together, for example water, H_2O, sodium chloride, NaCl or methane, CH_4.

The three types of bonding can now be considered in detail.

Ionic Bonding

Ionic bonds form when a metal transfers electrons to a non-metal, producing positive and negative ions.

When atoms transfer or gain electrons they obtain a more stable electron arrangement like their nearest **noble gas**. Consider the formation of the ionic compound, sodium chloride, from sodium and chlorine atoms. For these atoms to attain full outer shells of electrons the sodium atom (2,8,1) must transfer one electron to the chlorine atom (2,8,7). During this transfer process the sodium and chlorine atoms become charged and these charged atoms are known as **ions**. The positive sodium ion, Na^+ (2,8) is called a **cation** while the negative chloride ion, Cl^- (2,8,8) is called an **anion**. Figure 1 shows how these ions are formed.

Working out the charge on the ions for sodium chloride

From Table 1 it can be seen that:

- Before reacting, the chlorine and sodium atoms are neutral because they contain the same number of positive protons and negative electrons.
- After the atoms react and form ions, the charge on each ion is obtained by adding the number of protons and electrons; this gives the sodium ion a $^+1$ charge and the chloride ion a charge of $^-1$.

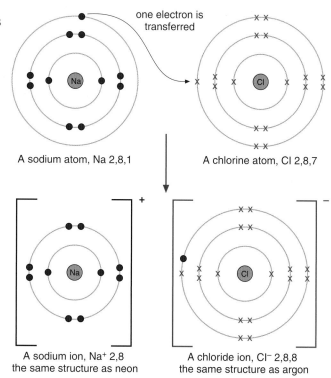

one electron is transferred

A sodium atom, Na 2,8,1 | A chlorine atom, Cl 2,8,7

A sodium ion, Na^+ 2,8 the same structure as neon | A chloride ion, Cl^- 2,8,8 the same structure as argon

The Na^+ ions and Cl^- ions are held together by electrostatic forces of attraction

Figure 1 Formation of the ionic compound, sodium chloride, NaCl

	Na atom	Na^{1+} ion	Cl atom	Cl^{1-} ion
Proton (+)	11(+)	11(+)	17(+)	17(+)
Electron (−)	11(−)	10(−)	17(−)	18(−)
Charge	0	1+	0	1−

Table 1 Calculating the charge on sodium and chloride ions in sodium chloride

An alternative method to work out the charge on ions is as follows:

● Metals will form positive ions while non-metals will form negative ions.

● The charge on the ions is equal to the number of electrons that are either lost or gained in forming a noble gas electron structure.

Applying these ideas to sodium chloride:

● Sodium is a metal with an electronic structure 2,8,1 and when it reacts with chlorine, it attains a noble gas electron structure by losing one electron. Thus it will form a positive ion (cation) with a charge of $^+1$.

● Chlorine is a non-metal with an electronic structure 2,8,7 and in its reaction with sodium, it gains one electron to attain a noble gas electron structure, forming a negative ion (anion) with a charge of $^-1$.

Electrostatic attraction

Since Na^+ ions and Cl^- ions have opposite charges there will be strong electrostatic attractions between these ions. The strong **electrostatic attractions** between ions are known as **ionic bonds**. In a crystal of sodium chloride there are many millions of oppositely charged ions bonded together in a giant three-dimensional ionic structure, as shown below in Figure 2. This giant ionic structure has the formula, Na^+Cl^- and is written as NaCl since the positive and negative charges on the ions cancel out. From the diagram it is seen that each Na^+ ion is surrounded by six Cl^- ions and each Cl^- is surrounded by six Na^+ ions.

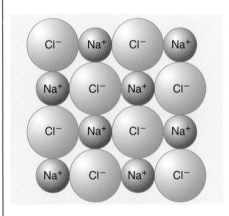 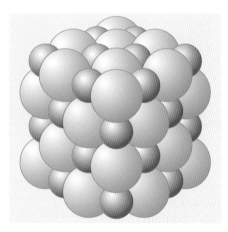

Figure 2 Sodium chloride crystal

Questions

1   Copy and complete the following passage for the reaction between potassium and fluorine:

Potassium and fluorine react to form potassium _____. When this happens each potassium atom transfers an _____ to a fluorine atom. As a result potassium ions with a _____ charge and fluoride ions with a _____ charge are produced. The ions are held together by strong _____ attractions. The chemical bond which results from this attraction is called an _____ bond.

2 a) Draw electron diagrams (dot and cross) to show the arrangement of the electrons in a lithium atom and a chlorine atom.

   b) How does the arrangement of the electrons in lithium and chlorine change when lithium chloride forms?

   c) Show how the charges on the lithium and chloride ions can be worked out.

   d) Which noble gases have the same electron structures as a lithium ion and a chloride ion?

3 Write out the electronic structures of the following atoms and use them to predict the charges that these atoms will have when they form ions:

   a) sodium   b) magnesium   c) calcium   d) aluminium
   e) oxygen   f) chlorine   g) sulphur

4 Use dot and cross diagrams to explain why sodium oxide has the formula Na_2O.

Ionic bonding in Group II metal compounds

To obtain a noble gas electron structure Group II metals must transfer 2 electrons when forming ionic compounds with non-metals. This is shown in Figure 3 for the formation of calcium chloride.

Physical properties of ionic compounds

From a knowledge of how ionic compounds are formed and how the ions are arranged in a regular crystalline lattice, it is possible to explain the physical properties of typical ionic compounds. Table 2 outlines the main physical properties of ionic compounds and provides an explanation for these properties.

Physical property	Reason
Ionic compounds are crystalline solids at room temperature, for example, salt (sodium chloride)	The positive and negative ions are packed closely together and are held by strong electrostatic attractions
Ionic compounds have high melting points, for example, aluminium oxide, Al_2O_3, 2045°C. Magnesium oxide, melting point 2800°C, is used for refractory bricks in the Blast Furnace	Large amounts of heat energy are required to separate the oppositely charged ions due to the strong electrostatic attractions (ionic bonds)
Ionic compounds conduct electricity when molten or dissolved in water but do not conduct in the solid state	When an ionic compound is melted or dissolved in water the strong electrostatic attractions are overcome and the ions are free to move and carry the current,(Figure 4). Ionic compounds cannot conduct electricity in the solid state as the ions are in fixed positions and are not free to move
Ionic compounds are generally soluble in water and insoluble in organic solvents, for example, sodium chloride dissolves in water to give a colourless solution but is insoluble in benzene	The water molecules are capable of colliding with the ionic solid, knocking off ions in the process. The ions and the water molecules then mix and diffuse evenly throughout the solution

Table 2 Properties of ionic compounds

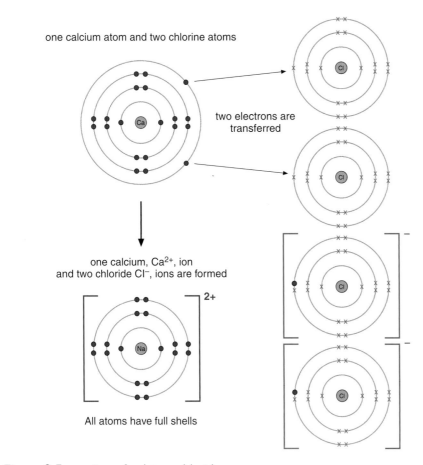

one calcium atom and two chlorine atoms

two electrons are transferred

one calcium, Ca^{2+}, ion and two chloride Cl^-, ions are formed

2+

All atoms have full shells

Figure 3 Formation of calcium chloride

Conduction of electricity by ionic compounds

Figure 4 shows that when ionic compounds are in the molten or dissolved state the negative anions move to the positive electrode (anode) while the positive cations move to the negative electrode (cathode). It is the ions which carry the current and the process is called **electrolysis**.

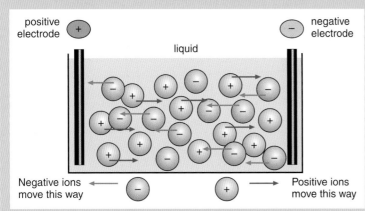

positive electrode

negative electrode

liquid

Negative ions move this way

Positive ions move this way

Figure 4 Ionic compounds conduct electricity when molten or dissolved

Questions

1 a) Draw diagrams to show the arrangements of the electrons in a magnesium atom and in a fluorine atom.

b) Use dot and cross diagrams to show how the arrangement of the electrons in a magnesium atom and a fluorine atom change when magnesium fluoride is formed.

c) Why is magnesium fluoride classified as a compound?

d) Give the formula of magnesium fluoride and state the type of bonding which exists in this compound.

2 Use your knowledge of ionic compounds to explain why:

a) Lead bromide conducts electricity when molten but not in the solid state

b) Magnesium oxide has a high melting point of 2800°C

c) Most ionic compounds are soluble in water.

3 The table below gives information about five different ions:

Ion	Number of protons	Number of neutrons	Number of electrons
A	12	12	10
B	19	20	18
C	8	8	10
D	11	12	10
E	9	10	10

a) Which two ions have a charge of 1+?

b) Select the correct formula for the compound formed between C and B
   CB C_2B B_2C C_2B_2

c) Give the charge on ion E

d) Use your periodic table to identify ion D.

4 **IT**: Produce a PowerPoint presentation showing the formation of sodium oxide. Your slides should include:

a) the electronic structure of sodium and oxygen

b) why the atoms react

c) the electronic structure of the ions

d) how the ions are held together

e) some typical properties of sodium oxide.

The study of ionic compounds shows how a chemical bond can be formed by transferring the outermost electrons from a metal to a non-metal in order that both types of atom can obtain a more stable electronic structure similar to that of their nearest noble gas. It is also possible for two non-metals to form a chemical bond by sharing electrons and this leads to another type of bonding known as covalent bonding.

Covalent bonding

Covalent bonding in elements

Covalent bonds are formed when two non-metal atoms share electrons to obtain a stable electron structure like their nearest noble gas. The simplest element, hydrogen, that has one electron in its outer shell forms a **covalent bond** when two hydrogen atoms share their electrons (Figure 5). When two electrons are shared in this way the bond is called a **single covalent bond**.

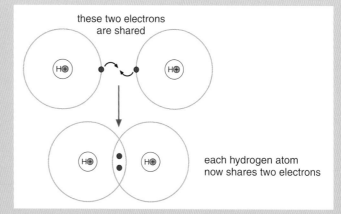

Figure 5 Formation of a hydrogen molecule, H_2

By sharing electrons each hydrogen atom has now obtained a noble gas electron structure like helium. The two atoms bonded together in this way are called a diatomic molecule, i.e. H_2. Here the term **diatomic molecule** refers to a small particle containing two atoms of the same element covalently bonded together.

Like hydrogen, chlorine forms a covalent diatomic molecule, Cl_2, where each chlorine atom, 2,8,7 shares an electron to obtain a noble gas electron structure like Ar 2,8,8. This is shown in Figure 6 below.

The chlorine molecule is sometimes written as Cl—Cl, where the line between the two chlorine atoms represents a single covalent bond

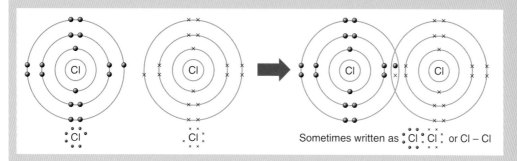

Figure 6 Formation of a chlorine molecule, Cl_2

There are cases when some non-metal elements must share more than one electron if they are to achieve a filled outer shell of electrons. An example is oxygen with an electronic structure of 2,6. Here oxygen needs to share two electrons to obtain a filled outer shell of electrons like neon 2,8 (Figure 7).

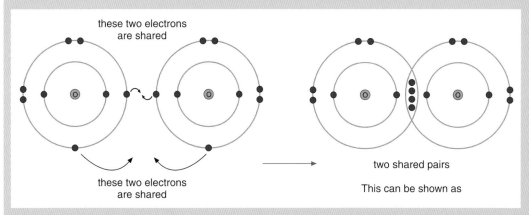

Figure 7 Formation of an oxygen molecule, O_2

Thus when two oxygen atoms combine they each share 2 electrons to achieve a filled outer shell of electrons and a double covalent bond is formed.

So far consideration has only been given to the formation of covalent bonds in molecules of the same element; it is also possible to form covalent bonds between different non-metallic elements. When this happens **covalent compounds** are formed.

Covalent bonding in compounds

Figure 8 shows hydrogen and oxygen combining to form the covalent compound water, H_2O. In this compound hydrogen and oxygen share electrons to achieve stable electronic structures like their nearest noble gases. Since oxygen has an electronic structure of 2,6 it will share with two electrons (one from each hydrogen atom) to obtain an electronic structure of 2,8. At the same time each hydrogen atom achieves an electronic structure similar to that of helium.

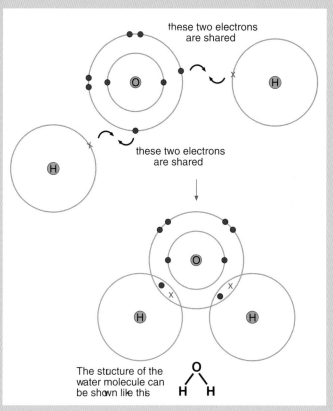

Figure 8 Formation of a water molecule, H_2O

Methane, CH_4 is a covalent gas formed when the two the non-metals, carbon and hydrogen, combine, (Figure 9). In this case, carbon with an electronic structure, 2,4 must share with four electrons from four hydrogen atoms for each of the elements to achieve a noble gas electronic structure.

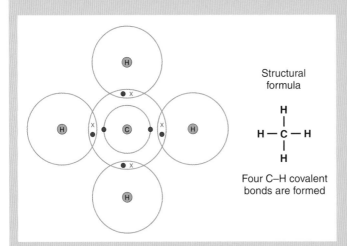

Figure 9 Covalent bonding in methane, CH_4

Structural formula

Four C–H covalent bonds are formed

Simple molecular substances

All the covalent substances discussed so far are composed of small molecules and are classified as **simple covalent molecular substances**. Such substances can be elements or compounds and they exist as solids, liquids or gases at room temperature. Although there are strong covalent bonds within the molecules there are only weak attractive forces between the molecules and this leads to simple molecular substances having low melting and boiling points (Table 3). Since the small molecules have no charges or free electrons they do not conduct electricity and it is generally found that they have low solubility in water.

Substance	State at room temperature	Formula of small covalent molecule	Structure of molecule	Melting point (°C)	Boiling point (°C)
Hydrogen	gas	H_2	H—H	−253	−259
Chlorine	gas	Cl_2	Cl—Cl	−34	−101
Bromine	liquid	Br_2	Br—Br	58	−7
Iodine	soft solid	I_2	I—I	184	114
Water	liquid	H_2O	O / \ H H	100	0
Methane	gas	CH_4	H ‖ H—C—H ‖ H	−162	−182

Table 3 Simple molecular substances and their physical properties

Questions

1 a) Explain what is meant by the term covalent bond.

 b) Using dot and cross diagrams show how the electrons are arranged in the following covalent molecules:

  (i) H_2 (ii) F_2 (iii) HCl   (iv) N_2 (v) CH_4 (vi) H_2O

2 a) Show how the electrons are arranged in

  (i) NH_3 (ii) CCl_4.

 b) Explain why the formula of ammonia is NH_3 and not NH_2.

 c) Although there are strong covalent bonds within the ammonia molecule, how can we explain that ammonia is a gas with a low boiling point of $-33°C$?

3 **IT:** Use PowerPoint to make a presentation showing the formation of methane, CH_4. Your slides should include:

 a) the electronic structures of carbon and hydrogen

 b) why the atoms react

 c) how the atoms are combined in the new compound

 d) some typical properties of methane.

Giant molecular substances

So far the discussion on covalent bonding has been based on simple covalent molecules. However, covalent bonding is also the bonding that holds giant molecular crystals together. Giant molecular structures exist for both elements and compounds and Figure 10 shows three examples, diamond, graphite and quartz (SiO_2).

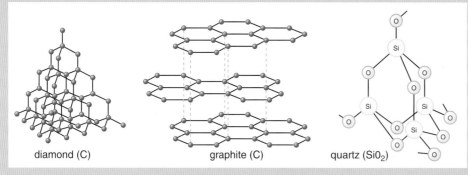

diamond (C)   graphite (C)   quartz (SiO_2)

Figure 10 Giant covalent structures of diamond, graphite and quartz

Diamond

From Figure 10 it is seen that each carbon atom in diamond is covalently bonded to four other carbon atoms in a giant covalent structure. The symmetrical, tetrahedral arrangement of carbon atoms along with the strong covalent bonds in all directions makes the structure very difficult to break apart.

The giant covalent structure gives diamond the following properties:

- one of the hardest naturally occurring substances
- melts at 3550°C and boils at 4827°C
- insoluble in water
- does not conduct electricity (there are no ions or free electrons to carry the current).

Due to diamond's great hardness and its beautiful sparkling appearance it has been used as a precious gemstone for over two thousand years. Diamond has also important uses as a glass cutter, in diamond-studded saws and in drill bits and this again results from its high strength and hardness.

Graphite

Like diamond, graphite is a naturally occurring form of carbon (Figure 10). Graphite is made of soft, greyish-black crystals which have a slippery feel. In graphite the carbon atoms are held by strong covalent bonds within the layers. The layers are made up of an interlocking system of hexagonal rings where each carbon atom is bonded to three others. This differs from the tetrahedral arrangement of carbon atoms in diamond, where each carbon atom is covalently bonded to four others. This means that each carbon atom in a layer of graphite has a free electron which is donated to a cloud of delocalised electrons and it is this cloud of free moving electrons which allows graphite to conduct electricity. While there is strong covalent bonding within in the layers, there are only weak attractive forces between the layers and so the layers of carbon can easily slide over one another. All of the following physical properties of graphite can be explained in terms of the strong covalent bonding within the layers and the weak attractive forces between the layers:

- high melting point, 3696°C
- high boiling point, 4827°C
- slippery solid with a soft feel
- conducts electricity
- insoluble in water

Because the layers in graphite can easily slip over one another it is used as a lubricant, for example, on bicycle chains and on metal parts in machinery. Mixed with clay and baked it is used to make pencil lead which can mark paper as the layers of graphite rub off and stick to the paper. Another important use of graphite, based on its electrical conductivity, is in the production of electrodes for electrolysis. In the commercial production of aluminium from the electrolysis of bauxite, graphite electrodes are used.

Quartz, silicon dioxide, SiO$_2$

Quartz, sand and flint are all forms of silica. From Figure 10 it is seen that the covalent structure of quartz is very similar to that of diamond where the silicon and oxygen atoms are joined by single covalent bonds in a giant three-dimensional pattern with an oxygen atom between every two silicon atoms. The giant tetrahedral arrangement of the silicon and oxygen atoms along with the strong covalent bonding between these atoms makes the structure difficult to break apart. This gives quartz the following physical properties:

- high melting point 1610°C
- high boiling point 2230°C
- hard solid
- insoluble in water
- non-conductor of electricity (there are no free electrons or ions to carry the current).

Optical glass which is made from quartz is used for the production of glass fibres in the telecommunications industry. Quartz crystals are also used to control the timing of the liquid crystal display in watches.

Questions

1 Make a list of the uses of diamond, graphite and quartz.

2 Make a table to show the similarities between diamond, graphite and quartz.

3 Diamond and graphite are both made of pure carbon but these two solids have very different properties and uses. In terms of their structures explain why:
   a) Diamond and graphite have very high melting points
   b) Diamond is an insulator but graphite is a good conductor of electricity
   c) Diamond is very hard and can be used in cutting drills
   d) Graphite is used as a lubricant.

4 In terms of its structure explain why:
   a) Glass bottles can be used to store concentrated acid
   b) Glass insulators can be used in electric pylons.

5 **IT**: Using Publisher design a poster to show the properties and uses of diamond. Your poster should include the structure of diamond and how the properties and uses are related to the structure.

Metallic Bonding

This occurs when metal atoms use their outermost electrons to form a *sea of electrons*, which move in all directions throughout the metallic structure as shown in Figure 11. These **free electrons** act like an 'electron glue' holding the metal ions together in a giant metallic structure. The attraction between the metal ions and the free moving electrons is known as the metallic bond.

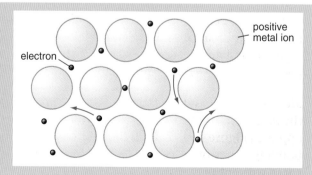

Figure 11 Metallic bonding

Metallic bonds are strong and account for the following physical properties of most metals:

● good conductors of heat and electricity
● high melting and boiling points
● bend easily and can be hammered into shape/malleable
● can be drawn into wires/ductile
● hard and have a high density.

Metals are good conductors of electricity and heat due to the sea of delocalised electrons. When a metal is connected in an electrical circuit the free electrons in the metal move towards the positive terminal. This movement of electrons towards the positive terminal causes an electric current to flow in the circuit as shown in Figure 12.

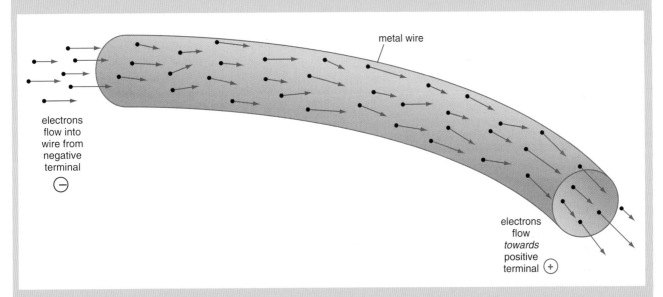

metal wire

electrons
flow into
wire from
negative
terminal
⊖

electrons
flow
towards
positive
terminal ⊕

Figure 12 Electrical conductivity of metals

Heating a metal causes the high-energy electrons in the 'sea of electrons' to move rapidly and randomly to cooler regions of the metal. During this process energy is transferred to all parts of the metal causing good conduction of heat. The high melting and boiling points of metals can be explained by considering the strong electrostatic attraction between the negative 'sea of electrons' and the positive metal ions. High energy is required to break the strong metallic bonds and this results in metals having high melting and boiling points. The strong metallic bond and the close packing of atoms also explain the high density of most metals.

Metals are malleable and ductile because the mobile 'sea of electrons' allows the atoms in the layers to move into different positions when a force is applied (Figure 13). This movement of electrons ensures that the metallic bonding is not broken and explains why metals can be bent or drawn into wires without breaking.

In Figure 13 it is seen that the sea of mobile electrons ensures that the close packed structure is continually restored when a force is applied.

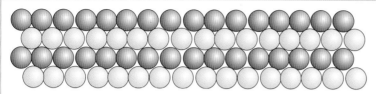

The atom arrangement in a straight piece of metal

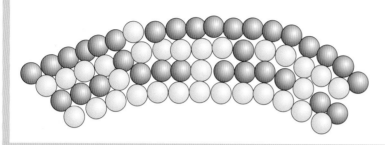

The atom arrangement when the metal is bent – some atoms are forced into different layers

Figure 13 The effects of bending a metal

Questions

1   Explain the meaning of the terms
a) malleable   b) ductile.

2   With the help of a diagram explain how the metallic bond forms in metals.

3   The properties of five substances are given in the table below.

Substance	Melting Point °C	Boiling Point °C	Electrical (solid)	Conductivity (liquid)
A	−101	−34	poor	poor
B	1083	2582	good	good
C	3550	4827	poor	poor
D	44	280	poor	poor
E	808	1465	poor	good

Use the letters A–E to answer the following questions and give reasons for your choices.
a) Which substance could be copper?
b) Which substance has a giant ionic structure?
c) Which two substances have simple molecular structures?
d) Which substance could be used in glass cutters?

4   Use your understanding of metallic bonding to explain why iron
a) is a good conductor of electricity
b) can be drawn into fine wires
c) is a fairly dense metal.

Websites

www.visionlearning.com/library/science/chemistry-1/ CHE1.7-bonding.htm

Exam questions

1 The table below contains information on five particles which are either atoms or ions.

Particle	Number of protons	Number of neutrons	Number of electrons
A	9	10	10
B	12	12	10
C	19	20	19
D	20	20	18
E	8	8	10

Use the letters, A, B, C, D or E to answer the following questions

a) Which particle has a charge of -2?
   (1 mark)

b) Which particle is a neutral atom?
   (1 mark)

c) Which particle is a magnesium ion?
   (1 mark)

2 When magnesium metal is burned in air it produces a white ash, magnesium oxide.

a) Draw diagrams to show how all the electrons are arranged in an atom of magnesium and in an atom of oxygen.

magnesium atom   oxygen atom

   (2 marks)

b) How does an atom of magnesium combine with an atom of oxygen to form magnesium oxide?
   (2 marks)

c) Name the type of bonding in magnesium oxide.
   (1 mark)

3 a) Draw a diagram to show the arrangement of **all** the electrons in a molecule of fluorine gas (F_2).
      (2 marks)

   b) What type of bonding would you expect for the compound tetrafluoromethane (CF_4)?
      (1 mark)

4 The element oxygen is made up of diatomic molecules.

   a) What do you understand by the term **molecule**?
      (2 marks)

   b) Using the **outer shells only** show how the electrons are arranged in an oxygen molecule.
      (2 marks)

   c) Explain why oxygen has a low boiling point.
      (2 marks)

5 The diagrams below show three high melting point solids.

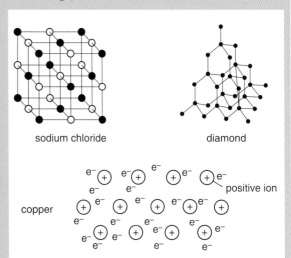

sodium chloride   diamond

copper   positive ion

a) Explain why sodium chloride can conduct electricity when molten but not when solid.

(2 marks)

b) (i) Explain why diamond is a very hard material.

(1 mark)

(ii) Give **one** use of diamond where its hardness is important.

(1 mark)

c) Explain why copper can be drawn into wire.

(1 mark)

6 a) The element calcium $^{40}_{20}$Ca, reacts with chlorine $^{35}_{17}$Cl, to form the compound calcium chloride.

(i) Give the meaning of the following words.

[A] Element *(2 marks)*

[B] Compound *(2 marks)*

(ii) Complete the following table for the two elements.

Element	Atomic number	Mass number	Number of protons	Number of neutrons	Electron arrange- ment
Calcium					
Chlorine					

(4 marks)

(iii) Describe the changes in electronic arrangements of the atoms of calcium and chlorine when they react. Charges on the ions should be included in your answer.

(6 marks)

(iv) Calcium chloride has a melting point of 782°C. Explain why the melting point is so high.

(2 marks)

b) The element chlorine contains atoms with two different mass numbers, ^{35}Cl and ^{37}Cl.

(i) Give the meaning of the term 'mass number'.

(2 marks)

(ii) How does the structure of a ^{35}Cl atom differ from that of a ^{37}Cl atom?

(2 marks)

(iii) What is the name given to atoms of the same element with different mass numbers?

(1 mark)

c) The table below gives the melting and boiling points of five substances.

Substance	Melting point/°C	Boiling point/°C
Chlorine	−101	−34
Zinc	420	913
Sodium chloride	808	1465
Quartz	1610	2230
Diamond	3550	4827

(i) What type of bonding is present in each of the substances in the table?

[A] Chlorine *(2 marks)*

[B] Zinc *(1 mark)*

[C] Sodium chloride *(1 mark)*

[D] Quartz *(2 marks)*

[E] Diamond *(2 marks)*

(ii) Why is the melting point of chlorine much lower than that of sodium chloride?

(4 marks)

Chapter 3

Elements, Compounds and Mixtures

Learning objectives

By the end of this chapter you should be able to:

➤ Classify substances as elements (metallic or non-metallic), compounds or mixtures

➤ Understand that compounds are substances that contain two or more elements chemically joined together

➤ Distinguish between elements, compounds and mixtures according to their properties

➤ Understand that mixtures are two or more substances that are usually easy to separate

➤ Understand that elements are substances that cannot be decomposed into simpler substances

➤ Understand how to write the formulae of elements and compounds

In our everyday life we are surrounded by millions of different substances, both naturally occurring and man-made. Fortunately for us, scientists have classified this vast range of substances into three groups: elements, compounds and mixtures. The basic building blocks for all these substances are elements and although there is an almost endless number of substances, they are all made from just over 90 naturally occurring elements.

Elements

Copper and limestone are examples of two common substances; limestone is not an element but copper is. We know that limestone is not an element because on heating it can be turned into simpler substances; however, it is not possible to do this for copper. When limestone is heated it decomposes into calcium oxide and carbon dioxide, but no matter how copper is treated it cannot be broken down further. Copper is an element and elements are defined as follows:

Elements are substances that cannot be decomposed into simpler substances.

Elements in the Periodic Table are classified according to their properties. The simplest way to classify elements is into metals and non-metals. In the Periodic Table over three-quarters of the elements are metals and the smaller number of the non-metals are positioned at the top right hand corner of the Table. The typical physical properties of metallic and non-metallic elements are summarised in Table 1. In Chapters 12 and 13 we will study more closely the properties of metals and of non-metals.

Figure 1 Some common everyday substances

Figure 2 Some common everyday elements and their uses

Metallic elements	Non-metallic elements
Good conductors of electricity	Poor conductors of electricity (graphite is an exception)
Good conductors of heat	Poor conductors of heat
All are solids at room temperature, except mercury	Can be gas, liquid or solid at room temperature
Usually have high density	Usually have low density
Usually have high melting and boiling points	Usually have low melting and boiling points
Usually malleable and ductile	Non-metallic solids tend to be soft and brittle (diamond is an exception)
Shiny solids	Usually dull (diamond is an exception)

Table 1 Typical physical properties of metals and non-metals

Probably the easiest way to show that an element is a metal or non-metal is to check out its electrical conductivity as shown in Figure 3. Metals will cause the bulb to light up because they are good conductors of electricity.

Table 2 provides information on the physical properties of elements and this allows us to make comparisons between the physical properties of metals and non-metals. It is seen that the metallic bonding in metals leads to high densities, high melting and boiling points while non-metals with a simple molecular covalent structure have low densities, low melting points and low boiling points. Non-metals that have a giant covalent structure have high densities, high melting points and high boiling points; in these respects they resemble metals.

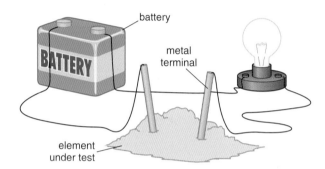

Figure 3 Testing for metals and non-metals

Element	Formula	Structure	Bonding	Melting point °C	Boiling point °C	Density g/cm³	State
Aluminium	Al	Giant metallic	Metallic	659	2447	2.7	Solid
Lead	Pb	Giant metallic	Metallic	328	1740	11.3	Solid
Tin	Sn	Giant metallic	Metallic	232	2260	7.3	Solid
Copper	Cu	Giant metallic	Metallic	1083	2582	9.0	Solid
Iron	Fe	Giant metallic	Metallic	1540	2887	7.9	Solid
Hydrogen	H_2	Simple molecular	Covalent	−259	−253	0.08*	Gas
Oxygen	O_2	Simple molecular	Covalent	−219	−183	1.33*	Gas
Nitrogen	N_2	Simple molecular	Covalent	−210	−196	1.17*	Gas
Bromine	Br_2	Simple molecular	Covalent	−7	58	3.1	Liquid
Sulphur	S_8	Simple molecular	Covalent	114	445	2.1	Solid
Diamond	C	Giant molecular	Covalent	3550	4827	3.51	Solid

(*Units are g/dm³ at 25°C, 1 atmosphere)

Table 2 How the physical properties of elements relate to their structure and bonding

From Table 2 it is seen that metallic elements always form giant metallic structures while non-metallic elements form simple covalent molecular structures or, in some cases, such as carbon, giant covalent molecular structures. It is the type of structure and bonding between the atoms in an element that determines its properties.

Writing chemical formulae for elements

Chemical formulae are a short-hand way of representing elements and compounds. When writing the formula of an element it is necessary to consider its structure. The following guidelines are helpful in writing an element's formula:

Elements with a giant metallic or giant covalent molecular structure

These elements have a formula which is just the same as their symbol, for example:

Element	Symbol	Formula
Diamond	C	C
Silicon	Si	Si
Copper	Cu	Cu
Sodium	Na	Na
Calcium	Ca	Ca
Iron	Fe	Fe

Gaseous and liquid elements with a simple covalent molecular structure

Apart from the noble gases the atoms of these elements form diatomic molecules and this is shown in the formula of their **molecules**; for example, a hydrogen molecule, H_2, contains two atoms in each small molecule.

Element	Symbol	Formula
Hydrogen	H	H_2
Oxygen	O	O_2
Nitrogen	N	N_2
Chlorine	Cl	Cl_2
Bromine	Br	Br_2

Element	Symbol	Formula
Helium	He	He
Neon	Ne	Ne
Argon	Ar	Ar

Solid elements with a simple covalent molecular structure

In GCSE the only two that you are likely to use are:

Element	Symbol	Formula
Iodine	I	I_2
Sulphur	S	S_8

Questions

1 Choose suitable elements to match the following three states

    a) a gas   b) a liquid   c) a solid.

2 Write symbols for the following elements

    a) potassium   b) barium   c) iron

    d) bromine   e) lithium.

3 Write the chemical formula for

    a) copper   b) silicon   c) oxygen gas   d) sulphur.

4 Which of the following substances is not an element

    a) sulphur   b) water   c) lead

    d) oxygen   e) silicon?

    Give a reason for your choice.

Compounds and mixtures

Water which makes up more than half the mass of the human body and covers around four-fifths of the world's surface is composed of oxygen and hydrogen chemically joined together. Water is described as a **compound** of hydrogen and oxygen. In the same way carbon dioxide is a compound and is composed of carbon and oxygen chemically joined together. Compounds are defined as follows:

Compounds are substances which contain two or more elements chemically joined together.

Table 3 shows a number of familiar compounds and the elements that combine to form them.

Common name	Chemical name	Formula	Elements present
Water	Hydrogen oxide	H_2O	Hydrogen and oxygen
Common salt	Sodium chloride	NaCl	Sodium and chlorine
Chalk	Calcium carbonate	$CaCO_3$	Calcium, carbon and oxygen
Lime	Calcium oxide	CaO	Calcium and oxygen
Sand	Silicon dioxide	SiO_2	Silicon and oxygen
Carbon dioxide	Carbon dioxide	CO_2	Carbon and oxygen
Sugar	Sucrose	$C_{12}H_{22}O_{11}$	Carbon, hydrogen and oxygen
Petrol	Octane	C_8H_{18}	Carbon and hydrogen
Alcohol	Ethanol	C_2H_5OH	Carbon, hydrogen and oxygen
Natural gas	Methane	CH_4	Carbon and hydrogen
Vinegar	Ethanoic acid	CH_3COOH	Carbon, hydrogen and oxygen

Table 3 Some common compounds

Looking at the differences between elements and compounds

To help us understand the differences between compounds and elements we can study the reaction between iron and sulphur that produces the compound iron sulphide.

$$iron + sulphur \rightarrow iron\ sulphide$$

When grey iron filings and yellow sulphur powder are added and mixed as shown in Figure 4:

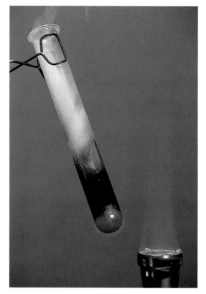

Figure 4 Iron and sulphur react to form iron sulphide

- The resulting mixture is intermediate in colour between that of the iron and sulphur.

- Using a magnifying glass it is possible to distinguish between the sulphur and iron particles.

- The iron and sulphur mixture can be made in any proportion.

- No heat change takes place when the elements are mixed.

- The mixture can be separated into its elements using a magnet; this attracts the iron filings from the mixture but does not attract the sulphur.

- When dilute sulphuric acid is added to the mixture it will react with the iron to produce hydrogen gas but the acid will not react with the sulphur and it remains undissolved in the acid solution.

We can say that the mixture shows the properties of the individual elements, iron and sulphur.

If the mixture is now heated:

- There is a chemical reaction and much heat is given out causing a bright red glow to pass through the mixture as the compound, iron sulphide, is formed.

$$iron\ + sulphur \rightarrow iron\ sulphide$$
$$Fe\ \ + S\ \ \ \ \ \ \ \rightarrow FeS$$

- The new compound is greyish-black in colour and it is impossible to distinguish between the particles of iron and sulphur.

- If the solid iron sulphide is ground into a powder it is now impossible to separate the iron from the sulphur using a magnet.

- The compound, iron sulphide, has completely different properties from those of either iron or sulphur; for example, when dilute sulphuric acid is added the solid disappears and the poisonous gas hydrogen sulphide is given off. In the case of the mixture it was hydrogen gas that was given off when sulphuric acid was added.

- The compound iron sulphide always contains a fixed mass of iron and sulphur combined in a given ratio.

From the above experiment it is possible to identify several differences between mixtures and compounds, Table 4 summarises the major differences:

Compound	Mixture
Is a pure substance	Contains two or more substances
A chemical reaction occurs when a compound is formed and there is an energy change	When a mixture is made, no chemical change takes place
The composition of a compound is always the same	The composition of a mixture can vary
The properties of a compound are different from those of the elements making up the compound	The properties of a mixture are the same as those of the individual elements making up the mixture
Chemical compounds are difficult to separate and are only broken down by chemical means; for example, electrolysis of molten sodium chloride gives sodium and chlorine	Mixtures are easy to separate using physical means; for example, sodium chloride crystals can be obtained from sodium chloride solution by evaporation

Table 4 Differences between compounds and mixtures

A mixture can now be defined as consisting of two or more substances that are usually easy to separate.

Some everyday examples of mixtures are given in Table 5.

Mixture	Type of mixture	Components
Rock salt	Solid/solid	Salt (sodium chloride) and sand
Sugar solution	Solid/liquid	Sugar and water
Whisky solution	Liquid/liquid (miscible liquids)	Alcohol and water
Soda water	Gas/liquid	Carbon dioxide and water
Air	Gas/gas	Mainly oxygen and nitrogen
Water/oil	Liquid/liquid (immiscible liquids)	Water and oil

Table 5 Some everyday examples of mixtures

Questions

1   Using examples, explain the meaning of the following terms:

   a) element   b) compound   c) mixture.

2   Separate the following substances into elements, compounds and mixtures:

   copper, marble chips, milk, baking soda, oxygen, sugar, carbonated water, air, sulphur, salt, alumina, diet coke, carbon dioxide and seawater.

3 Complete the following word equations to show how elements combine to form compounds:

magnesium + oxygen →
calcium + sulphur →
sodium + chlorine →
sulphur + oxygen →
hydrogen + chlorine →

4 a) Here is a list of formulae

H_2^+ H H^+ H^{2+} H_2

Which of the formulae represents:
  (i) a Hydrogen molecule?
  (ii) a Hydrogen ion?

b) Ozone gas has the formula O_3. Is ozone an element, compound or mixture?

5 Use the following website to create a database of the first 20 elements of the Periodic Table:

  http://www.chemsoc.org

a) Include the following information in the database: formula, metal or non-metal, density, boiling point and melting point.

b) Set up queries to determine those elements which are: solids at 293K, liquids at 293K and gases at 293K.

c) What is the melting point of the most dense metal?

d) What is the boiling point of the most dense non-metal?

6 Using the above website select a metal, a liquid non-metal and a gaseous element from the Periodic Table and make a poster by saving and printing the information on each element's discovery, appearance, sources and uses.

Names and chemical formulae

Names of ionic compounds

The formula for an ionic or a giant covalent compound is the whole number ratio in which the ions or atoms exist in the compound. For example, the formula of the ionic compound, calcium chloride is $CaCl_2$, this tells us that there are two chloride ions for every calcium ion in the compound. For the giant covalent compound, silica, SiO_2, there are two oxygen atoms for every silicon atom. For the simple molecular compound, CH_4, there are four atoms of hydrogen and one atom of carbon in each molecule.

The name of a chemical compound tells us the elements present; for example, magnesium oxide tells us that magnesium and oxygen are present. In the compound's name it is seen that the name of the metal is always given as the first part of the name, while the non-metal's name appears as the second part of the name. To show it is a compound and not a mixture the non-metal part usually changes its ending to **–ide** as shown in Table 6.

Reacting elements	Compound produced
Magnesium + ox**ygen**	Magnesium ox**ide**
Iron + sulph**ur**	Iron sulph**ide**
Sodium + fluor**ine**	Sodium fluor**ide**
Sodium + chlor**ine**	Sodium chlor**ide**
Potassium + brom**ine**	Potassium brom**ide**
Sodium + iod**ine**	Sodium iod**ide**
Magnesium + nitr**ogen**	Magnesium nitr**ide**

Table 6 How elements change their names when they combine

Some ionic compounds end in **–ate** or **–ite**; both these endings indicate that the ionic compound contains oxygen as well as the metal and the non-metal. For example, sodium sulphate, Na_2SO_4, contains sodium, sulphur and oxygen while sodium nitrite, $NaNO_2$, contains sodium, nitrogen and oxygen. The compound sodium carbonate (Na_2CO_3) contains sodium, carbon and oxygen.

Formulae of ionic compounds

The charges on the different cations and anions as given in Table 7 will be used to work out the formula of ionic compounds. The charge on an ion is also related to the **element's valency** or **combining power**. The sodium with a charge of 1+ has a valency of 1 while the aluminium with a charge of 3+ has a valency of 3.

Cation 1+	Symbol	Cation 2+	Symbol	Cation 3+	Symbol
Lithium	Li^{1+}	Magnesium	Mg^{2+}	Aluminium	Al^{3+}
Sodium	Na^{1+}	Calcium	Ca^{2+}	Iron(III)	Fe^{3+}
Potassium	K^{1+}	Barium	Ba^{2+}		
Ammonium	NH_4^{1+}	Copper(II)	Cu^{2+}		
Silver	Ag^{1+}	Zinc	Zn^{2+}		
Hydrogen	H^{1+}	Lead	Pb^{2+}		
		Iron(II)	Fe^{2+}		

Anion 1−	Symbol	Anion 2−	Symbol	Anion 3−	Symbol
Fluoride	F^{1-}	Carbonate	CO_3^{2-}	Phosphate	PO_4^{3-}
Chloride	Cl^{1-}	Sulphate	SO_4^{2-}	Nitride	N^{3-}
Bromide	Br^{1-}	Sulphide	S^{2-}		
Iodide	I^{1-}	Oxide	O^{2-}		
Hydroxide	OH^{1-}				
Hydrogencarbonate	HCO_3^{1-}				
Hydrogensulphate	HSO_4^{1-}				
Nitrate	NO_3^{1-}				

Table 7 Valencies or ion charges for a number of cations and anions

When writing the formula of an ionic compound the following rules must be used:

- Write down the cation and anion present in the compound.
- Because ionic compounds are neutral, the charges on the cations and anions must cancel out.

The following examples show how to apply the above rules:

- Sodium chloride: ions present are Na^{1+} and Cl^{1-}; because $(1+) + (1-) = 0$ and cancel out, the formula is NaCl.
- Calcium oxide: ions present are Ca^{2+} and O^{2-}; because $(2+)$ and $(2-)$ cancel each other out, the formula is CaO.
- Sodium sulphide: ions present are Na^{1+} and S^{2-}, in this example it takes two Na^{1+} to cancel out the 2− charge on S^{2-}; $2 \times (1+) + (2-) = 0$, the formula is Na_2S.
- Calcium bromide: ions present Ca^{2+} and Br^{1-}, in this example it takes two Br^{1-} to cancel out the 2+ charge on the Ca^{2+}; $(2+) + 2 \times (1-) = 0$, the formula is $CaBr_2$.
- Aluminium oxide: ions present Al^{3+} and O^{2-}, in this example it takes two Al^{3+} and three O^{2-} to provide a neutral compound; $2 \times (3+) + 3 \times (2-) = 0$, the formula of the compound is Al_2O_3.
- Calcium nitrate: the ions present are Ca^{2+} and NO_3^{1-}, in this example it takes two NO_3^{1-} to cancel out the 2+ charge on Ca^{2+}; $(2+) + 2 \times (1-) = 0$. The formula in this compound is written as $Ca(NO_3)_2$. Brackets are used around the nitrate anion because it is a group of atoms.

Formulae of covalent compounds

The following table provides information on the covalent compounds covered at GCSE. Some of the formulae below were studied under covalent bonding (pages 17–22).

Compound	Formula	Solid, liquid or gas
Carbon dioxide	CO_2	Gas
Carbon monoxide	CO	Gas
Ammonia	NH_3	Gas
Hydrogen chloride	HCl	Gas
Methane	CH_4	Gas
Nitrogen dioxide	NO_2	Gas
Sulphur dioxide	SO_2	Gas
Water	H_2O	Liquid
Sulphuric acid	H_2SO_4	Liquid
Nitric acid	HNO_3	Liquid
Hydrogen peroxide	H_2O_2	Liquid
Sulphur trioxide	SO_3	Solid
Quartz	SiO_2	Solid

Table 8 Some common covalent compounds

A number of the formulae but not all can be worked out by considering valencies or combining power of each atom in the molecules. The following examples show how the formula are obtained:

- Ammonia, NH_3: nitrogen has a valency of 3 while hydrogen has a valency of 1. This means that it requires 3 hydrogen atoms to combine with 1 nitrogen so the formula is NH_3.

- Methane, CH_4: carbon has a valency of 4 while hydrogen has a valency of 1. This means that it takes 4 hydrogen atoms to combine with 1 carbon atom so the formula is CH_4.

- Silica, SiO_2: silicon has a valency of 4 while oxygen has a valency of 2. This means that it takes 2 oxygen atoms to combine with 1 silicon atom.

The above study of elements, compounds and mixtures provides us with the background knowledge and understanding to study the next chapter on the different classes of materials and their everyday uses.

Questions

1 Using calcium chloride as an example explain the meaning of the term valency.

2 Write formulae for the following ionic compounds.

a) potassium bromide
b) sodium oxide
c) magnesium chloride
d) barium nitrate
e) lithium carbonate
f) copper(II) hydroxide
g) calcium hydrogencarbonate
h) iron(II) sulphate
i) lead chloride
j) aluminium sulphate.

3 Write formulae for the covalent molecules below.

a) carbon dioxide
b) water
c) silicon chloride
d) ammonia
e) methane
f) hydrogen chloride

4 Write out the formulae for the cations and anions in the following compounds.

a) Na_2SO_4
b) $Mg(NO_3)_2$
c) $PbSO_4$
d) $BaCl_2$
e) $ZnCO_3$

Websites

www.webelements.co.uk

Exam questions

1 Each of the boxes in column A can be matched with a box in column B below. Copy the diagram. Draw lines to match the chemicals in column A to the statements in column B. One has been done for you.

Column A   Column B

Column A	Column B
iron and sulphur	contains two elements which are chemically combined
sulphur	conducts heat and electricity
iron sulphide	can be separated into two elements using a magnet
iron	a yellow solid with a pungent smell

(3 marks)

2 Chemicals can be identified by symbols, formulae or names. Five examples of chemicals found in school laboratories are given below.

C   MgO   Sodium chloride   Mg $CuSO_4$

You may use your Data Leaflet to help you answer this question.

a) Give the **name** of a chemical from the list which is an element.

(1 mark)

b) Write down the formula for sodium chloride

(1 mark)

c) What is the **name** of the last chemical labelled $CuSO_4$?

(1 mark)

3 Four chemicals are labelled as shown below. You may find your Data Leaflet useful in answering this question.

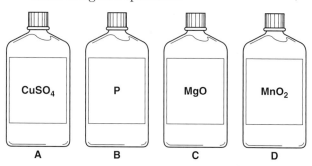

a) What is the **name** of the element which has the symbol P?

(1 mark)

b) Which bottle, **A**, **B**, **C** or **D**, contains magnesium oxide?

(1 mark)

c) Why can magnesium oxide be described as a **compound**?

(2 marks)

4 The diagrams below show how particles are arranged in five different substances.

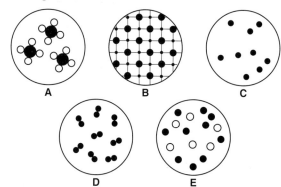

Use the letters **A, B, C, D** and **E** to answer the following questions.

Which diagram shows:

a) an element which is made up of single atoms?

(1 mark)

b) a mixture of more than one element?

(1 mark)

c) a diatomic gas? *(1 mark)*

d) an ionic solid? *(1 mark)*

e) molecules of methane? *(1 mark)*

5 The symbol equation below represents the thermal decomposition of calcium carbonate.

$$CaCO_3 \rightarrow CaO + CO_2$$

a) Write this equation out again as a **word** equation.

Calcium carbonate → Calcium + Carbon Oxide Dioxide.   *(2 marks)*

The substance in the above equation are all **compounds**.

b) What is a compound?   *(2 marks)*

c) How could the thermal decomposition of calcium carbonate be carried out in the laboratory?

(1 mark)

6 Copy and complete the table below to show the correct formulae for the compounds listed **and** the total number of atoms present in each formula.

Name	Formula	Number of atoms
Calcium chloride	$CaCl_2$	Add all Symbols etc
Ammonium nitrate	NH_4HNO_3	
Potassium carbonate	K_2CO_3	
Magnesium hydroxide	$MgOH_2$	
Calcium hydrogencarbonate	$Ca(HCO_3)_2$	
Aluminium sulphate	$AlSO_4$	

(12 marks)

7 The table below gives some information about the melting points and boiling points of five chemicals, A, B, C, D and E.

Chemical	Melting point (°C)	Boiling point (°C)
A	72	216
B	−15	58
C	1304	2176
D	−142	−63
E	795	1147

Which chemical, A, B, C, D or E:

a) is a liquid at room temperature (20°C)?

(1 mark)

b) could melt if placed in boiling water?

(1 mark)

c) has the lowest boiling point?   *(1 mark)*

d) would diffuse most easily?   *(1 mark)*

8 The diagrams below show the electronic structures of the atoms of three elements. You may find your data leaflet useful in answering this question.

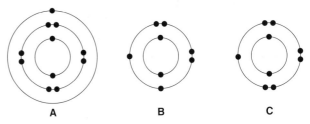

A   B   C

a) Which **one** of the atoms, **A, B** or **C** could form an **ion** with a charge of −1?

(1 mark)

b) Draw the electronic structure of the **ion** produced from atom **A**.

(1 mark)

c) Draw a circle around the correct formula of the compound formed by atoms **A** and **B**.

(i)   **A₂B**

(ii) **AB**

(iii) **AB₂**

(1 mark)

9 Some properties of elements are given in the table below.

Element	Boiling Point (°C)	Electrical Conductivity	Density (g cm³)
A	765	Good	8.65
B	−152	Poor	0.0034
C	445	Poor	2.07
D	2970	Good	19.3
E	58	Poor	3.12
F	1330	Good	0.53

Use the letters A–F to answer the following questions.

a) Which element is a gas at room temperature?

(1 mark)

b) Which is the least dense metal?

(1 mark)

c) Give **two** elements which are made up of small molecules.

(1 mark)

Chapter 4

Materials and Their Uses

Learning objectives

By the end of this chapter you should be able to:

➤ Recognise and know the value of common hazard symbols on containers, i.e. flammable, toxic, corrosive, explosive and harmful/irritant

➤ Relate your knowledge of the properties of the different classes of man-made materials to everyday use. Such materials include metals, ceramics, glass, fibres and plastics (thermosoftening and thermosetting)

➤ Relate the properties of thermosoftening plastics, thermosetting plastics and fibres to simple models of their structure

➤ Understand that a composite material is one that combines the properties of more than one material to produce a more useful material for particular purposes

➤ Evaluate the relative advantages and disadvantages of composite materials e.g. glass fibre (boats and car bodies), reinforced glass (windows), reinforced concrete (beams) and bone (skeleton)

Note: Chapter 4 is only relevant to double award foundation students and double award higher students.

Scientists use the term 'material' to describe the different types of matter that are used to make things. For example glass or plastic can be used to make bottles, copper is used to make pipes, while paper is used in the production of bags. In everyday life there are a wide range of materials available and these materials can be used for many different purposes.

Because there is a continuous demand for new materials with specialised properties, it is important for scientists to design new products to meet these needs. An example of this is the special ceramic coatings which have been designed for the outside of space shuttles to protect them from burning up on

a)

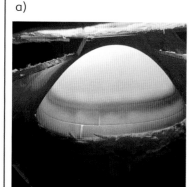

b)

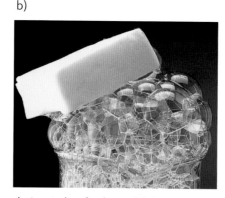

c)

Figure 1 a) The nose cone of a shuttle is made of a 'reusable' carbon composite, b) SEAgel is the first lighter-than-air solid and is produced from agar, it could be used as a thermal insulator and c) Supergel is a water-based hydrogel capable of absorbing up to half its own weight in oil.

re-entry into the earth's atmosphere. Another example is kevlar, probably the most important man-made fibre ever developed. It is five times stronger than steel, but is not as heavy and does not corrode. Kevlar is used to make anchor lines for ships and bulletproof vests. Pyrex glass is used in the laboratory and in cooking instead of ordinary soda-lime glass. Ordinary glass cracks under a sudden change in temperature but scientists have modified its properties by adding boron oxide to produce pyrex glass. This type of glass doesn't crack when the temperature is suddenly changed.

Different kinds of materials

Materials can be classified as natural and man-made materials, for example, wood, silk and cotton are natural materials while iron, glass, ceramics, nylon and polythene are man-made. A more useful way of classifying materials is by their properties. Using this method of classification leads to five major groups; **metals**, **ceramics**, **glass**, **plastics** and **fibres**.

Figure 2 Materials in the home

In Figure 2 it is seen that:

● The metal cans are made from aluminium. Aluminium, like most metals is extracted from its ore, bauxite, by electrolysis.

● The glasses are made from glass and although there are many different types of glass they are all made by heating sand, limestone and sodium carbonate at a temperature of 1300 to 1400°C. A thick viscous liquid is produced that can be worked and blown into the required shape. As the glass cools down it becomes hard and brittle. Scientists term glass a supercooled liquid rather than a solid because the atoms have a random arrangement similar to that of a liquid. Glass has an irregular giant covalent structure of silicon and oxygen (page 22) that contains trapped calcium and sodium ions.

● The china plates are made from **ceramics**. The word ceramic is derived from the Greek word 'keramos' which means pottery. Ceramics have a similar structure to glass and are made from a mixture of sand and clay. This mixture of sand and clay is moulded into the desired shape, dried and then heated or 'fired' in an oven at around 1000°C. A hard, insoluble, glassy ceramic is produced. Ceramics can be glazed by coating the moulded clay with a suitable glaze mixture before firing.

● The jeans is made from cotton which is a natural material and consists of long, thin thread like strands that are very strong. Cotton, wool, linen, polyester, nylon and rayon are classified as **fibres** and, like polythene, are polymers with a giant covalent structure. Natural fibres are those which are made by living things, for example, silk, linen, cotton and cellulose while synthetic fibres are man-made and examples include polyester, nylon and acrylics.

● The **plastic** bottles have been made from the polymer, polythene. Ethene is obtained from crude oil and is made into polyethene by polymerisation. It is called a synthetic polymer because it is man-made. A polymer is a large molecule that is formed by chemically bonding together many small molecules called monomers. Polythene has a giant covalent structure and while there are strong covalent bonds between the atoms in the chains, there are only weak attractive forces between the chains. This covalent structure explains the physical properties of polythene. In our study of plastics two types will be considered, thermosoftening and thermosetting.

Thermosoftening and thermosetting plastics: properties

Thermosoftening plastics are flexible and can be moulded into different shapes. This is explained by the fact that the long chains in the polymer can:

● stretch easily
● soften on warming
● are flexible
● can be shaped by warming.

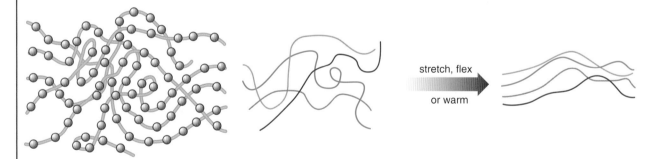

stretch, flex

or warm

Figure 3a Tangled chains of long, thin molecules in a thermosoftening plastic

Figure 3b The effect of heat on a thermosoftening polymer

On the other hand, **thermosetting** plastics are strong, rigid materials and once formed they do not soften or melt and cannot be remoulded. As can be seen in Figures 4.4a and 4.4b the long chain molecules are joined by cross-links and when the chains are heated the crosslinks between the chains stop them from moving over each other. As a result, this type of plastic cannot be melted or remoulded and, if heated, will burn or char. The cross-linking between the chains forms a rigid network and this results in the plastic being hard, unlike thermosoftening plastics which are flexible due to their long chains which have no cross-linking.

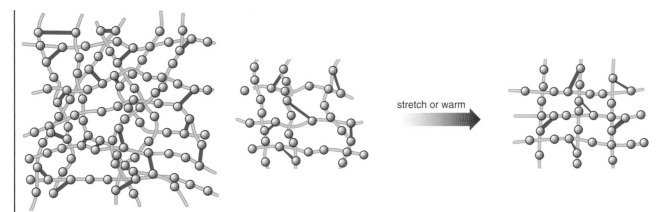

Figure 4a Cross-links in a thermosetting plastic

Figure 4b The effect of heat on a thermosetting polymer

stretch or warm

Thermosetting and thermosoftening plastics: uses

Thermosetting and thermosoftening plastics have giant covalent molecular structures and can act as insulators because the electrons are fixed in strong covalent bonds. Table 1 gives several other important physical properties of thermosoftening and thermosetting polymers.

Since the nineteen sixties many new plastics have been discovered and these have replaced traditional materials such as wood, metal, glass, paper, ceramics, silk, wool and cotton. The properties of these new plastics make them suitable for a wide range of uses; for example, they are cheap, have a low density, are unreactive, and are easy to dye. In consequence they are used to make boats, cars, toys, packaging material, clothing, and other such items. A major disadvantage of synthetic plastics is the difficulty of their disposal as they are non-biodegradable (are not broken down by bacteria in the soil). Recent developments have seen the introduction of new biodegradable and photodegradable plastics that could eventually replace the present ones. Cost and government legislation on the use and disposal of plastics will determine the extent to which there is a demand for these new materials.

The properties of the five classes of materials are listed in Table 1. These properties enable scientists to make decisions regarding which material is most suitable for a given purpose.

Type of material	Examples	General properties	Manufacturing process
Metals	Iron, aluminium, lead copper and zinc	<ul><li>Hard</li><li>Strong</li><li>Malleable</li><li>Good conductors of heat and electricity</li><li>Ductile</li><li>High density</li><li>High melting and boiling points</li><li>Can react with acids, water and air</li></ul>	Made from metal ores. Iron is obtained from the ore haematite, Fe_2O_3 while copper is obtained from the ore copperpyrites, $CuFeS_2$

Table 1 The five classes of material

Type of material	Examples	General properties	Manufacturing process
Glass	Soda glass (bottles and windows) Pyrex (heat resistant glass)	● Brittle ● Hard ● Very unreactive ● High melting point ● Strong in compression and low tensile strength ● Transparent	Made by heating sand, limestone and sodium carbonate at 1300 to 1400°C
Ceramics	Pottery, bricks, china, tiles and crockery	● Brittle ● Hard ● Very unreactive ● High melting point ● Strong in compression but low tensile strength	Mainly made from clay heated to around 1000°C
Fibres	Polyester, lycra, acrylics, nylon, cotton, linen and silk	● Flexible ● Low density ● Long, thin strands of fibre ● Burns on heating	Synthetic fibres are manufactured from crude oil, while natural fibres are obtained from plants and animals
Plastics Thermosoftening	Polythene, polypropene, PVC and polystyrene	● Flexible ● Easy to melt and remould ● Insulators ● Low density ● Many burn on heating	Synthetic plastics manufactured from crude oil
Thermosetting	Melamine, bakelite and epoxy resins	● Strong and stiff ● Insulators ● Low density ● Cannot be remoulded ● Decompose, burn or char when heated	Manufactured from crude oil

Table 1 The five classes of material – *continued*

Properties of materials

Some important properties of materials are listed below.

Hardness

This property of materials relates to how easy it is to scratch or dent a material. A harder material will always scratch a softer material. The Mohs scale is used to compare the relative hardness of materials and ranges from 1 for talc to 10 for diamond.

Brittle

A brittle material such as glass will shatter if dropped on a hard floor. Brittle materials such as glass or ceramics cannot absorb the energy of a large collision without cracking or shattering totally.

Strength

A strong material is one which is difficult to break when a force is applied. Strength is usually associated with a stretching force (strength in tension). For example, it is possible to investigate the stretching force required to break a nylon thread. Materials that are difficult to break by stretching, are said to have high tensile strength. Metals generally display high tensile strength. Strength can also be considered in terms of compression which is the effect of crushing or squeezing a material. Glass and ceramics are strong when compressed but weak when stretched. They are said to have good compressive strength.

Flexibility

A flexible material is one that is easy to bend without breaking. Flexible materials have good tensile strength and compressive strength, because on bending one side of the material is stretched while the other is compressed. Plastics and fibres are examples of flexible materials.

Hazardous chemicals

Many chemicals or materials have harmful properties and their containers must be labelled with hazard symbols to identify the dangers. Hazard symbols are used because they have been internationally agreed and are easily recognised. Words beside or below the symbols indicate the type of hazard. Figure 5 shows diagrams of the different hazard symbols used to identify harmful chemical substances:

Explosive

A substance likely to explode.
e.g. ammonium dichromate

Toxic

Substances which can cause death. They may cause problems if swallowed, inhaled or absorbed through the skin.
e.g. chlorine, sulphur dioxide

Harmful

Substances similar to toxic substances but not as harmful
e.g. aspirin, iodine, ethanol

Corrosive

Substances which attack and destroy living tissues, including eyes and skin.
e.g. concentrated sulphuric acid, concentrated sodium hydroxide

Irritant

Substances which are not corrosive but can cause reddening or blistering of the skin
e.g. dilute sulphuric acid, iodine solution

Highly flammable

Substances which can catch fire easily
e.g. petrol, ethanol

Figure 5 Hazard symbols

Questions

1   Use the website http://www.creative-chemistry.org.uk/gcse/module 7.htm and answer the questions on hazard symbols on the worksheet 'Using Hazard Symbols'.

Uses of materials

Scientists choose a material for a particular function because the material's properties are suited to that use. For example, marble is used as a building material because it is hard and strong, while polythene is used to make washing up bowls because it is waterproof and easy to mould. When choosing a material for a particular purpose there are a number of questions that must be asked. Consider the problem of deciding on a suitable material for making disposable coffee cups at a football stadium. In making this decision we are likely to ask the following questions:

Are the physical properties suitable? By this we mean properties such as density, flexibility, melting point, strength, brittleness and conductivity.

Are the chemical properties suitable? Here we would consider chemical properties related to reactivity in air, water, acid and/or alkali.

Is the cost of the material suitable? The cost of the material must not be too high or the product will be too expensive.

The following reasons could be used in deciding why polystyrene is the most suitable material for disposable cups.

Physical properties: easily moulded, lightweight, insulator and, because it will not be reused, it is sufficiently strong.

Chemical properties: a very unreactive material that will not react with the drink, water or air.

Cost of material: readily available and cheap to manufacture.

The table below highlights some of the uses for the five types of materials. In each case we can see how the use is related to a material's properties.

Material	Use	Property
Iron	Bridges	High strength
Aluminium	Saucepans	Good conductor of heat
Copper	Electrical wiring	Good conductor of electricity
Lead	Roofing	Malleable
Zinc	Galvanising iron	Chemically more reactive than iron
Ceramic tiles	Flooring	Very hard
Soda glass	Windows	Transparent
Pyrex glass	Glassware for cooking	Resistant to changes in temperature
PVC	Clothes	Waterproof and flexible
Polythene	Bottles	Easy to melt and mould
Polystyrene	Wall insulation	Good insulator
Polyester	Clothes	Strong and flexible
Melamine (thermosetting)	Kitchen worktops	High heat resistance
Bakelite (thermosetting)	Electric plugs	Insulator and does not melt
Epoxy resins (thermosetting)	Glues	Excellent adhesive properties

Table 2 Some uses of material

Composite Materials

It is possible to design a material that combines the desirable properties of two different materials. In sport, many tennis racquets, fishing rods and golf club shafts are made from carbon-fibre reinforced plastic. This type of material provides the flexibility of plastics and the very high strength of carbon fibres. In most cases this material has replaced the more traditional material, wood. Carbon-fibre reinforced plastic is an example of a composite material.

Figure 6 Composite materials in sport

Composite materials are those which combine the properties of more than one material and produce a more useful material for a particular purpose. In a composite material one of the materials acts as a matrix while the other is used as a filler. This is shown in Figure 7 for glass-fibre reinforced plastic, where the plastic acts as the matrix and holds the filler, glass fibre, in place.

As well as being man-made or synthetic there are many composite materials found in nature, for example ivory, teeth and bone. Bone combines the properties of calcium phosphate with those of long chain proteins. Calcium phosphate is required to make the bones hard, while the long chain protein makes it more flexible than calcium phosphate on its own.

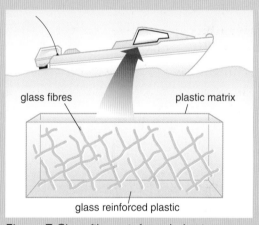

Figure 7 Glass-fibre reinforced plastic

Four common composite materials are shown in Figure 8 and their relative advantages in the table on the next page.

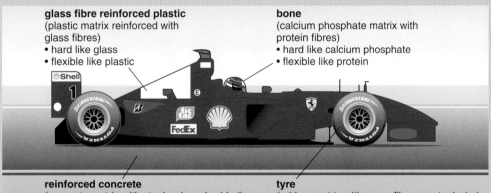

glass fibre reinforced plastic
(plastic matrix reinforced with glass fibres)
• hard like glass
• flexible like plastic

bone
(calcium phosphate matrix with protein fibres)
• hard like calcium phosphate
• flexible like protein

reinforced concrete
(concrete matrix with steel rods embedded)
• cheap and hard like concrete
• strong like steel

tyre
(rubber matrix with rayon fibres or steel wire)
• flexible and elastic like rubber
• strong like rayon or steel

Figure 8 Some composite materials

Composite	Materials present	Advantages	Uses
Reinforced glass	Glass matrix reinforced with wire mesh	Stronger and less brittle than glass	Windows and doors
Reinforced concrete	Concrete matrix reinforced with steel rods	Hard like concrete with the flexibility and strength of steel	Construction industry, for example bridges and buildings
Glass-reinforced plastic	Plastic matrix reinforced with glass fibre matting	Flexible like plastic. Much stronger than plastic and less brittle than glass	Boats, car bodies and canoes
Bone	Calcium phosphate reinforced with protein fibres	Hard like calcium phosphate but more flexible than calcium phosphate	Skeleton

Table 3 Relative advantages of four composites materials

Questions

2 a) Give one example of a composite material in nature and one example of a man-made composite that is used in the construction industry.

   b) What are the advantages of using a composite material instead of a single material?

3

Material	Density (kg/m³)	Relative strength	Relative stiffness	Cost
Steel	7800	1	210	Low
Polythene	960	0.02	0.6	Low
Kevlar	1450	3	190	High
Carbon Fibre Reinforced Plastic	1600	1.8	200	High
Nylon	1100	0.08	3	Medium

   a) Use the information in the table to give one advantage of using steel for car bodies.

   b) Use the information in the table to give two reasons why Kevlar is used for making bulletproof vests.

   c) Carbon fibre reinforced plastic can be described as a composite material. What do you understand by the term composite material.

Websites

www.gcsechemistry.com/ukop.htm

www.sciencenet.org.uk/database/Chemistry/Materials/C00133b.html

Exam questions

1 The label below is found on a bottle of bleach.

WARNING

Do not mix with other products. may release dangerous gases. (Chlorine)

Keep out of reach of children

a) From the information on the label, say why the bleach must be kept out of the reach of children.

(1 mark)

b) Why are hazard symbols used on labels?

(1 mark)

c) This label is found on a bottle of white spirits, a **flammable** solvent.

Copy and complete the label by drawing in the correct hazard symbol for a flammable substance.

WARNING

Not to be taken internally

Keep out of reach of children

(1 mark)

2 a) The following table shows how some plastics are used in many houses.

Copy and complete the table by giving **two** properties of each plastic and explain why it is used in the way stated.

(8 marks)

Plastic	Use	Property
polystyrene	roof insulation	
PVC	plastic guttering	
polythene	plastic bucket	
melamine	kitchen work top	

b) (i) Plastics can be classified as thermosoftening or thermosetting.

Explain precisely what is meant by these two terms.

(4 marks)

(ii) Using only those plastics in the table in part a) give **one** example of a thermosoftening plastic and **one** example of a thermosetting plastic.

An example of a thermosoftening plastic is

(1 mark)

An example of a thermosetting plastic is

(1 mark)

c) (i) Many plastic articles, e.g. polythene bags create litter problems because they are non-biodegradable.

What do you think the term 'non-biodegradable' means?

(2 marks)

(ii) Suggest **two** methods which would help minimise the litter problems plastic articles create.

(2 marks)

3 The table below gives information about some metals.

Metal	Density (g/cm³)	Melting point °C	Tensile strength (relative)	Corrosion behaviour	Cost per tonne
Aluminium	2.7	659	2	Corrodes very slowly	£750
Copper	9.0	1083	4	Corrodes very slowly	£1000
Iron	7.9	1540	4	Corrodes (rusts) quickly	£130
Lead	11.3	328	1	Corrodes very slowly	£290
Silver	10.5	961	3	Corrodes slightly (tarnishes on surface)	£150,000

Using the information:

a) Give **one disadvantage** of using aluminium for tent poles instead of iron.
(*1 mark*)

b) State which metal would be **most** suitable for boots worn by deep sea divers, and give a reason for your choice.
(*2 marks*)

4 a) What do you understand by the term composite material?
(*2 marks*)

Tyres are made from rubber and steel.

b) Give **one** disadvantage for each of these materials if they were to be used alone.

(i) Rubber:
(*1 mark*)

(ii) Steel:
(*1 mark*)

5 **Carbon-fibre reinforced** plastic is a material which is being considered for use in replacement surgery.

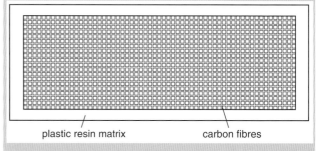

plastic resin matrix   carbon fibres

a) What type of material is carbon fibre reinforced plastic?
(*1 mark*)

Material	Density	Relative strength	Relative stiffness	Cost
Steel	7800	1	210	Low
Polythene	960	0.02	0.6	Low
Kevlar	1450	3	190	High
Carbon-fibre reinforced plastic	1600	1.8	200	High
Nylon	1100	0.08	3	Medium

b) From the table above write down **one advantage** and **one disadvantage** of using carbon-fibre reinforced plastic for replacement surgery instead of steel.
(*2 marks*)

Chapter 5

Matter and the Kinetic Theory

Learning objectives

By the end of this chapter you should be able to:

➤ Explain changes of state (including sublimation), diffusion and dissolving in terms of the kinetic theory and the energy changes associated with them

➤ Understand the terms solvent, solute, solution, saturated solution, hydrated and dehydration

➤ Recognise the factors affecting solution, i.e. heating, stirring, surface area and volume of solvent

➤ Understand the qualitative effect of temperature on the solubility of solids and gases in water

➤ Carry out simple quantitative determination of solubility of solids in water leading to an understanding of solubility curves

➤ Recognise that gases have weight and that they spread out to fill the space available, for example diffusion of bromine

➤ Give examples to show that gases are compressible and how this property is used in everyday life, for example, fire extinguishers and aerosol sprays

➤ Recognise that the volume of a gas depends upon pressure and temperature

➤ Use the relationship between the volume of a gas and its pressure and temperature to solve simple problems, i.e. PV/T = constant (conversion to STP and quantitative practical details are not required)

Matter

Materials can be classified into solids, liquids and gases and these are referred to as the three states of matter. The important physical properties of the three states of matter are summarised in Table 1.

Kinetic Theory

Scientists use the kinetic theory to explain the behaviour of matter. It is used to explain many familiar things that happen in everyday life, for example why:

● Does water freeze when it is cooled?

● Do metals expand on heating?

● Can we smell food cooking in a restaurant?

Property	Solids	Liquids	Gases
Shape	Stay the same shape	Flow easily Take the shape of their container	Flow easily Take the shape of their container
Volume	Stay the same volume	Stay the same volume	Take the volume of their container
Density	High densities (usually greater than 2 g/cm³)	Medium densities (about 1 g/cm³)	Low densities
Compressibility	Cannot be compressed	Can be compressed very slightly	Can be compressed into a much smaller volume
	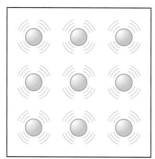	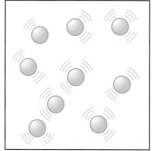	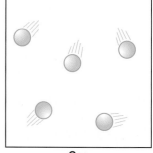

Table 1 The important properties of solids, liquids and gases

The kinetic theory states that:

● All matter is made up of tiny, invisible moving particles. These particles can be atoms, ions or molecules.

● The particles are continually moving and if the temperature is raised the particles gain energy and move more quickly.

● At a given temperature lighter particles move more quickly than heavier ones do.

In Figure 1 the kinetic theory is used to explain the differences between solids, liquids and gases:

Solid
The particles are close together, and are held in place by strong forces between them. So solids are more dense than liquids and gases and cannot be compressed. The particles can only vibrate about fixed points. So solids have a fixed volume and shape and cannot flow.

Liquid
The particles are a little further apart and the forces between them are not as strong as those in solids. So liquids are not as dense as solids and can be compressed slightly. The particles can roll around each other so they flow easily. They can change their shape but have a fixed volume.

Gas
The particles are very far apart and there are no forces between them. So the densities of gases are very low and they can be easily compressed. The particles move around very fast in all the space available, so they fill their whole container and flow easily.,

Figure 1 Particles in solids, liquids and gases

Using the kinetic theory to explain changes of state

The different changes of state are shown in Figure 2.

Melting

When a solid is heated, the particles gain energy causing them to vibrate faster and faster about their fixed positions. When sufficient energy is gained the particles break away from their fixed positions and the solid melts and a liquid forms. The particles can now move around each other as shown in Figure 3. The temperature at which this occurs is called the melting point of the solid. During the melting process heat energy is taken in by the particles and this energy is used to overcome the bonds or the attractions between the particles. Melting is described as an **endothermic** process because energy has been taken in to melt the solid. Solids with high melting points have strong bonds between the particles and a lot of energy must be supplied to break these strong bonds.

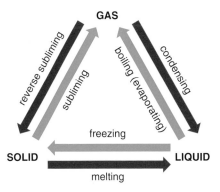

Figure 2 Changes of state

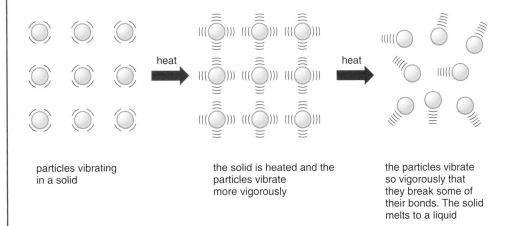

Figure 3 What happens when a solid melts

particles vibrating in a solid

the solid is heated and the particles vibrate more vigorously

the particles vibrate so vigorously that they break some of their bonds. The solid melts to a liquid

Boiling and evaporation

When a liquid is heated the particles gain energy and move around faster. Before reaching the boiling temperature, some of the higher energy particles will have enough energy to overcome the attractions between themselves and other particles in the liquid. These particles are able to escape and form a gas. This is known as **evaporation**. On further heating all particles gain sufficient energy to overcome the attractive forces between particles and the liquid boils. Due to the large number of particles escaping at the boiling point, bubbles of vapour are formed throughout the liquid. Liquids with strong attractive forces between the particles have high boiling points while those with weaker attractive forces have lower boiling points. Like melting, boiling and evaporation are endothermic processes, where energy is absorbed or taken in.

While melting, evaporating and boiling take in energy, **condensation** and **freezing** give out energy i.e. they are exothermic processes. Consider cooling a gas to form a liquid. During cooling, the gas particles lose energy and are able to move closer and closer together until condensation takes place. The formation of attractive forces between the particles in the liquid causes energy to be given out. Similarly, when the particles of a liquid are cooled, the particles move closer and closer together until the attractive forces between the particles are strong enough for freezing to take place. The bonds, which form when a liquid freezes to form a solid, cause energy to be given out.

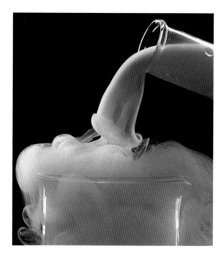

Sublimation

Sublimation occurs when a solid goes directly from a solid to a gas. Cooling the gas causes it to go directly back to the solid state. This is an unusual change because on heating the solid or cooling the gas the substance does not pass through the liquid state. Solid carbon dioxide (dry ice) and iodine crystals are two examples of substances that sublime. Dry ice is a solid at $-79°C$ and above this temperature it sublimes to form carbon dioxide gas. Solid carbon dioxide is a useful coolant and can be used to create special effects on stage. Iodine is a grey-black solid at room temperature and when it is gently heated it produces a deep purple vapour. Figure 4 shows the effect of heating both dry ice and iodine.

Figure 4 Dry ice and iodine subliming

Taking the kinetic theory further

The kinetic theory can be used to explain what happens during the processes of dissolving and diffusion and how pressure or temperature affects the volume of a gas.

Dissolving

We are familiar with solids such as salt dissolving in water to form a colourless solution. Solids such as salt are **soluble** in water and the resulting mixture is called a salt **solution**. The salt is called the **solute**, while water is called the **solvent**.

Solute + solvent → solution
Salt   + water   → salt solution

The time taken to dissolve a given amount of salt in water can be decreased by (i) increasing the volume of water (ii) increasing the temperature of the water (iii) stirring the mixture or by (iv) increasing the surface area of the salt. When a solid dissolves in water the particles which make up the solid become evenly distributed throughout the water molecules and the process is shown in Figure 5.

When sugar dissolves in water the moving water molecules crash into the sugar particles causing them to break away from the crystal and into the water. The sugar and water particles then mix and diffuse evenly throughout.

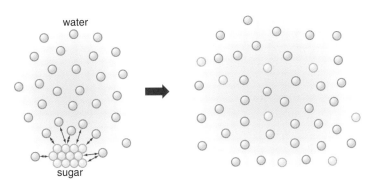

water

sugar

the water particles bombard the
sugar particles...

...and pull them apart so that they
spread out in the water

Figure 5 What
happens when a sugar
crystal dissolves in
water.

When a gas dissolves in water the molecules of the gas are distributed evenly
throughout the solution in a similar way to the particles of a solid dissolved in
water.

Questions

1   What do you understand by the following terms:

   a) solute   b) solvent   c) solution

2   Name the processes involved in the following changes of state:

   a) a solid changes to a liquid   d) liquid changes to a gas
   b) a solid changes to a gas   e) liquid changes to a solid
   c) steam changes to water   f) gas changes to solid

3   For each part of Question 2 state if the change taking place gives out or
takes in energy.

4   Use diagrams to explain what happens to
   a) the particles of a gas as it is cooled down to a liquid and then cooled
further to a solid
   b) the particles of iodine when it sublimes

5   Use the kinetic theory to explain what happens when
   a) a crystal of salt dissolves in water
   b) iron metal is heated and it expands
   c) water evaporates

6   **IT:** Searching the web.
   Use the website www.dryiceinfo.com to obtain three uses of dry
ice. Outline the advantages of using dry ice for the uses you have
chosen.

Dissolving and solubility

Consider when a small amount of copper(II) sulphate is added to water at
room temperature, the copper(II) sulphate dissolves and the solution is said to
be unsaturated. If more and more of the compound is added at this
temperature, a stage is reached when no more copper(II) sulphate will be
dissolved and some of it remains undissolved in the solution. We say that the
copper(II) sulphate solution is **saturated**. A saturated solution is one that
contains as much of the dissolved salt as possible at a given temperature.

It is possible to calculate how many grams of a solute are required to form a saturated solution. This is called the **solubility of the solute**, which is defined as the number of grams of solute that can dissolve in 100 g of water at a given temperature.

Different salts have different solubilities; for example, at 20°C the solubility of sodium nitrate is 88 grams/100 grams water and this is much greater than copper(II) sulphate which is 21 grams/100 grams water. The solubility of a substance can be calculated experimentally using the procedures outlined in Figure 6. The experimental results have been obtained for potassium chlorate.

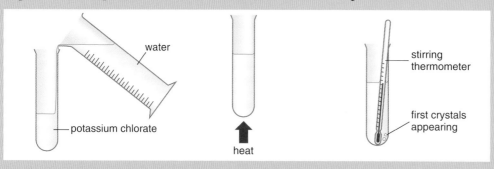

Figure 6
Determining the solubility of potassium chlorate

1   Place 2 grams of potassium chlorate in a boiling tube and add 4 cm³ water
2   Heat the boiling tube gently and stir until the potassium chlorate is just dissolved
3   Stirring, allow the solution to cool and record the temperature at which the crystals first appear (see table below)
4   Now add a further 4 cm³ water to the boiling tube and repeat steps 2 and 3.
5   Repeat steps 2 to 4, each time adding 4 cm³ water until the total volume of water is 20 cm³ (see table below)

Volume of water (cm³)	Temperature at which potassium chlorate crystals form (°C)	Solubility (grams of potassium chlorate/100 g water)
4	93	2 g potassium chlorate in 4 g water at 93°C $2 \times \dfrac{100}{4} = \textbf{50 g}$ potassium chlorate/100 g water
8	63	2 g potassium chlorate in 8 g water at 63°C $2 \times \dfrac{100}{8} = \textbf{25 g}$ potassium chlorate/100 g water
12	49	2 g potassium chlorate in 12 g water at 49°C $2 \times \dfrac{100}{12} = \textbf{16.67 g}$ potassium chlorate/100 g water
16	36	2 g potassium chlorate in 16 g water at 36°C $2 \times \dfrac{100}{16} = \textbf{12.5 g}$ potassium chlorate/100 g water
20	28	2 g potassium chlorate in 20 g water at 28°C $2 \times \dfrac{100}{20} = \textbf{10.0 g}$ potassium chlorate/100 g water

Table 2 Measuring the solubility of potassium chlorate

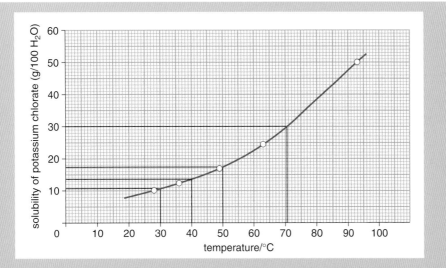

Figure 7 Solubility curve for potassium chlorate

The results for the solubility of potassium chlorate at different temperatures can now be used to plot a **solubility curve**, as in Figure 7.

The solubility curve allows us to work out:

● *The solubility of potassium chlorate at a given temperature in the range 27°C to 93°C.*
For example, the solubility at 40°C is 13.7 g/100 g water.

● *The temperature at which a potassium chlorate solution will become saturated and produce crystals*
For example, a solution containing 30 g/100 g water will start to form crystals at 70.5°C.

● *The mass of potassium chlorate crystals which will form when a saturated solution of potassium chlorate is cooled to a lower temperature*
For example, if a saturated solution of potassium chlorate in 100 g water at 50°C is cooled to 30°C, the mass of crystals formed is 17.3 − 10.7 = 6.6 g.

To compare the solubilities of different substances it is useful to plot their solubility curves. The solubility curves for a number of compounds are shown in Figure 8.

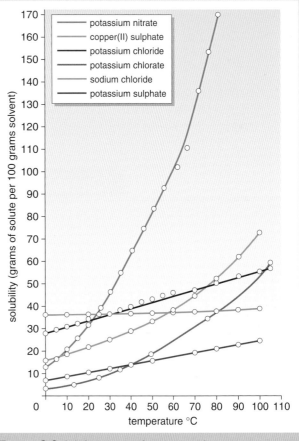

potassium nitrate
copper(II) sulphate
potassium chloride
potassium chlorate
sodium chloride
potassium sulphate

Figure 8 Solubility curves for some substances in water

From the curves it is seen that the solubility of solids increases with increasing temperature. It is also seen that the solubility of some substances increases more than others as the temperature is increased. However, there are some compounds such as sodium chloride where the solubility remains fairly constant as the temperature increases.

Unlike solids, the solubility of gases in water decreases as the temperature increases. This has important consequences for fish and other aquatic life. Fish need oxygen and they absorb the gas from water through their gills. At 0°C the solubility of oxygen is 4.8 cm^3/100 cm^3 water while at 20°C, the solubility decreases to 3.3 cm^3/100 cm^3. An increase of 20°C causes water to lose approximately 30 per cent of its dissolved oxygen. In hot summer days when temperatures are high vital supplies of oxygen are lost from the water and this can be a threat to the fish population and other aquatic life.

Hydration and dehydration

When some solids crystallise from aqueous solution they contain water and are said to be **hydrated**. When blue hydrated copper(II) sulphate crystals are heated in a test tube, water vapour is given off and the blue crystals change to a white powder called anhydrous copper(II) sulphate. The water that is given off is chemically combined with the copper sulphate and forms part of the compound, hydrated copper sulphate, $CuSO_4.5H_2O$. Water chemically joined in this way is called **water of crystallisation**. We can summarise the change taking place by the equation:

$$CuSO_4.5H_2O \qquad CuSO_4 \qquad\qquad + 5H_2O$$

blue hydrated	$\rightarrow$	white anhydrous
copper(II) sulphate		copper(II) sulphate + water

When hydrated copper(II) sulphate loses its water of crystallisation it is said to be dehydrated. **Dehydration** is described as the removal of the elements of water from a compound to form a new compound.

Questions

1   Explain the meaning of the following terms:

a) saturated solution   b) solubility of a solid   c) solubility curve

2

Temperature (°C)	0	10	20	30	40	50	60
Solubility of potassium chloride/g per 100 g water	28	31	33	36	39	42	45
Solubility of potassium nitrate/g per 100 g water	13	21	31	47	63	83	106

The above table gives the solubilities of two potassium salts.

a) Draw solubility curves for the two potassium salts. Plot solubility per 100 g of water on the vertical axis and temperature on the horizontal axis.

b) Calculate the solubility of each salt at 47°C.

c) How much potassium chloride will crystallise when 100 grams of water saturated with this salt at 56°C is cooled to 32°C.

3 The solubility of sodium nitrate in water is given below:

a) Draw the solubility curve for sodium nitrate by plotting solubility

Temperature (°C)	10	20	30	40	50	60	70
Solubility of sodium nitrate (g/100 g H₂O)	80	88	96	105	114	124	135

on the vertical axis and temperature on the horizontal axis

b) Calculate the solubility of sodium nitrate at (i) 35°C and (ii) 64°C

c) How much sodium nitrate will crystallise if a saturated solution of sodium nitrate in 100 g water is cooled from 64°C to 35°C?

d) At what temperature is the solubility of sodium nitrate (i) 91 g/100 g water and (ii) 120 g/100 g water?

e) State whether the following solutions of sodium nitrate in water give saturated solutions:

(i) 90 g sodium nitrate in 100 g water at 35°C

(ii) 118 g sodium nitrate in 100 g water at 47°C

(iii) 121 g sodium nitrate in 100 g water at 68°C.

4 The following results were obtained in an experiment to calculate the solubility of potassium chlorate at 27°C:

mass of evaporating basin = 20.25 g

mass of evaporating basin + saturated solution at 27°C = 31.25 g

mass of evaporating basin + solid potassium chlorate after evaporation to dryness = 21.25 g

a) calculate the solubility in g/100 g water of potassium chlorate at 27°C

b) what is the mass of a saturated solution containing 25 g of water?

Diffusion

We are familiar with diffusion in everyday life; for example, we can smell food cooking in the kitchen or smell perfume in a room. In both cases it does not take long for the smell to travel from one room to the next. The food and the perfume give off particles of gas which then move or diffuse throughout the rooms. **Diffusion** provides us with evidence that the particles of a gas can move and it is described as the movement of particles from a region of high concentration to a region where the concentration is less.

As gas particles move freely it is not surprising that they fill up any space which is available to them and this can be demonstrated using bromine vapour as shown in Figure 9. The red/brown bromine particles diffuse quickly into the air particles and at the same time, the air particles diffuse into the red/brown bromine particles. Eventually the colour in each gas jar is the same as the air and bromine particles have completely mixed.

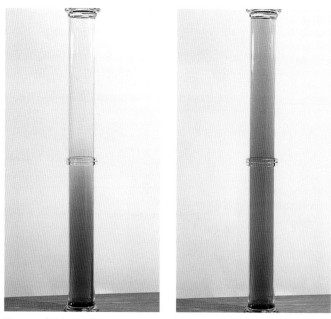

Figure 9 The diffusion of bromine

Diffusion in liquids is much slower that that in gases because the particles in a liquid move much more slowly than those in a gas. This can be demonstrated by allowing a purple crystal of potassium permanganate to diffuse in a beaker of water. The diffusion of the crystal can take a few days to complete and the process is shown in Figure 10.

At first the purple potassium permanganate crystal starts to dissolve in the water and there is a high concentration of the potassium permanganate at the bottom of the beaker. As time progresses both the water and potassium permanganate particles move and mix and eventually the purple colour is spread evenly throughout the solution when diffusion is complete.

Diffusion can also be used to show that different gases can diffuse at different speeds. The speed that a gas can move depends on the mass of the particles of the gas. For example, ammonia gas, which is a lighter gas than hydrogen chloride, travels at a faster speed. This can easily be shown by the experiment shown in Figure 11.

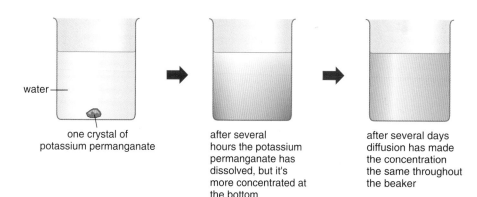

water

one crystal of potassium permanganate

after several hours the potassium permanganate has dissolved, but it's more concentrated at the bottom

after several days diffusion has made the concentration the same throughout the beaker

Figure 10 The diffusion of potassium permanganate in water

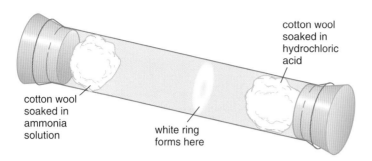

Figure 11 The diffusion of ammonia and hydrogen chloride

cotton wool soaked in hydrochloric acid

cotton wool soaked in ammonia solution

white ring forms here

The ammonia and hydrogen chloride molecules diffuse or move along the tube and meet, producing a white ring of ammonium chloride.

Ammonia + hydrogen ⟶ ammonium
              chloride   chloride
                                (white solid)

At first we might expect the ring to form in the middle of the tube. However, the white ring forms closer to the hydrogen chloride end as the lighter molecules of ammonia travel faster than the slower moving molecules of hydrogen chloride. The experiment clearly shows that lighter particles diffuse much faster than heavier particles.

Effect of pressure and temperature on a gas

Earlier we found out that the particles in a gas are far apart and move freely with little or no attraction between them. If pressure is applied to a gas at a given temperature then it is possible to push the particles closer together. Figure 12 shows how this can be easily demonstrated using a gas syringe. If you hold your finger over the outlet at the bottom of the syringe and push down on the barrel of the syringe, you will find that the volume of the gas decreases as the pressure is increased. This can be explained by saying that the gas molecules in the syringe can be pushed closer together as the pressure is increased and as a result the volume of the gas decreases. By the same argument the volume of a gas at a given temperature increases when the pressure is decreased.

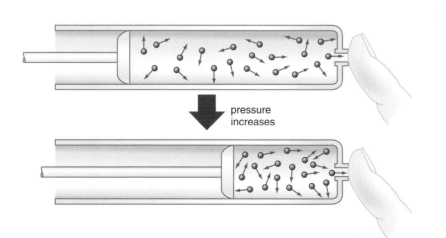

pressure increases

Figure 12 The effect of pressure on the volume of a gas

The fact that gases are compressible is important in the production and design of aerosol sprays and fire extinguishers. In aerosols, the gas which acts as the propellant is stored under pressure and when the valve is opened the propellant gas is released carrying with it the active component, for example hair lacquer, shaving foam or paint. In carbon dioxide fire extinguishers, dilute acid and sodium hydrogencarbonate react to produce carbon dioxide under pressure. The large pressure set up forces a froth of carbon dioxide to pass through the jet and over the flame. The blanket of carbon dioxide and water extinguish the flame.

If the temperature of a gas is increased at a given pressure then the volume of the gas increases. This can be easily shown using the apparatus in Figure 13. As the person's hands heat the flask, the air inside gets warmer and expands. This happens as the particles gain energy, move faster and further apart. We know that the air has expanded because we can observe air bubbling through the water.

If the temperature is decreased at a given pressure then the volume will also decrease. Cooling the round bottomed flask causes the water level to rise in the glass tubing because there is a decrease in the volume of air.

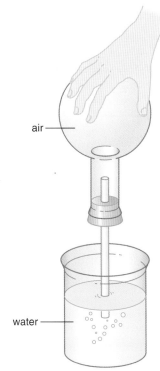

air

water

Figure 13 How temperature affects the volume of a gas at constant pressure

Mathematical relationship between volume, pressure and temperature of a gas

The relationship between the volume of a gas and its pressure and temperature was investigated by Robert Boyle in 1662 and by Jacques Charles in 1787. Their findings were published as Boyle's law and Charles' Law and have been combined into a single equation known as the **gas equation**:

$$PV/T = \text{constant} \quad \text{or} \quad P_1V_1/T_1 = P_2V_2/T_2$$

Where:

P_1, V_1, T_1 and P_2, V_2, T_2 correspond to the pressure, volume and temperature of a gas under two different sets of conditions.

Pressure is measured in **pascals** (Pa) or newtons per square metre (N/m^2). As the pascal is a very small unit, pressure is often given in kilopascals kPa.

Temperature in the Gas Equation must be given in **Kelvin** (K). To convert a Celsius temperature (°C) to Kelvin (K) you add 273 to the number of degrees Celsius, for example 25°C = (273 + 25) giving 298 K

Volume is measured in cubic metres (m^3) but you may use other units, for example cubic centimetres (cm^3) or cubic decimetres (dm^3).

The following example shows how to use the gas equation to calculate the new volume which a gas occupies when the pressure and temperature are changed.

Worked Example
80 cm^3 of oxygen were placed in a gas syringe at 27°C and a pressure of
12 000 Pa. The gas was then heated to 87°C at a new pressure 18 000 Pa.
What is the volume of oxygen under the new conditions?

First write out P, V and T for each set of conditions;

$P_1 = 12\,000$ Pa $\qquad\qquad$ $P_2 = 18\,000$ Pa
$V_1 = 80$ cm^3 $\qquad\qquad\quad$ $V_2 = ?$ cm^3
$T_1 = (273 + 27) = 300$ K $\qquad$ $T_2 = (273 + 87) = 360$ K

Substitute these values into the gas equation, $P_1V_1/T_1 = P_2V_2/T_2$, and
calculate the new volume V_2 in cm^3.

$$\frac{(12\,000 \times 80)}{300} = \frac{(18\,000 \times V_2)}{360}$$

Making V_2 the subject of the formula we get:

$$V_2 = \frac{(12\,000 \times 80 \times 360)}{(300 \times 18\,000)}$$

$$V_2 = 64 \text{ cm}^3$$

We see that under the new conditions of temperature and pressure the
volume has decreased from 80 cm^3 to 64 cm^3.

Questions

6 Explain the meaning of the following terms:

   a) diffusion $\qquad\qquad$ b) compressible $\qquad\qquad$ c) kinetic theory

7 Use your understanding of the kinetic theory to explain why:

   a) when a bottle of perfume is opened at the back of a classroom, a short
       time later it is possible to smell the scent at the front of the room
   b) gases diffuse much faster than liquids
   c) sugar dissolves faster in hot water than in cold water.

8 A bicycle pump contains 75 cm^3 of air at a temperature of 300 K
and a pressure of 120 000 Pa (120 kPa). Use the equation
PV/T = constant to calculate the volume of air in the pump if the
temperature is increased to 320 K and the pressure is increased to
150 000 Pa (150 kPa).

9 A gas syringe holds 64 cm^3 of oxygen gas at a pressure of 10 000 Pa
and a temperature of 300 K.

   a) What would the volume of the gas be if the pressure was
       doubled to 20 000 Pa and the temperature kept at 300 K?
   b) What would the volume of gas be if the temperature was doubled
       to 600 K and the pressure was kept at 10 000 Pa?
   c) What would the volume of gas be if the temperature was doubled
       to 600 k and the pressure was doubled to 20 000 Pa

Websites

www.dryiceinfo.com

Exam questions

1 a) The diagram below shows a burning candle.

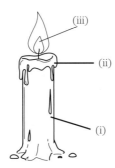

Copy the diagram and in the space, draw the arrangements of the particles at positions (i), (ii) and (iii).

(3 marks)

b) It is possible to interchange the states of matter. The following diagram shows these changes.

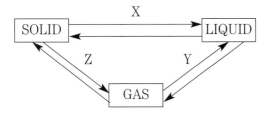

(i) Name the changes X, Y and Z.

(3 marks)

(ii) Which of the changes X, Y or Z is achieved by a decrease in temperature?

(1 mark)

c) Soft solder is a mixture of lead and tin. Heat is given out when it changes from liquid to solid. Explain this in terms of particle theory.

(2 marks)

d) Explain the essential difference between the **evaporation** of water and the **decomposition** of water.

(2 marks)

e) The following apparatus was set up to investigate the movement of two gases.

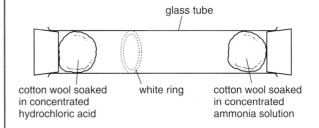

(i) What name is given to the movement of gases?

(1 mark)

(ii) Name the gas given off by concentrated hydrochloric acid.

(1 mark)

(iii) Explain fully why the white ring was formed closer to the concentrated hydrochloric acid end of the tube.

(2 marks)

f) A gas syringe contains 80 cm³ of gas at 280 K and 1 atmosphere pressure. What pressure would this same amount of gas exert if the volume was decreased to 40 cm³ and the temperature increased to 350 K?

(2 marks)

2 a) Which word best describes each of the following changes?

    (i) Turning a solid into a liquid.

        (1 mark)

    (ii) Turning a solid into a gas.

        (1 mark)

    (iii) Turning a gas into a liquid.

        (1 mark)

    (iv) Turning a liquid into a solid.

        (1 mark)

    (v) Turning a liquid into a gas.

        (1 mark)

b) A gas jar was filled with bromine gas and a second (empty) gas jar was inverted on top of the first jar, as shown in the diagram.

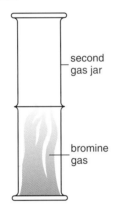

second gas jar

bromine gas

    (i) Describe what would be observed after the jars were in contact for some time.

        (2 marks)

    (ii) Explain your observations in part b)(i) in terms of particles.

        (2 marks)

    (iii) What name is given to the observation in part b)(i)?

        (1 mark)

c) A small crystal of potassium permanganate was added to a large beaker of water. After some time a purple colour had spread through the water.

Explain this observation in terms of particles.

        (3 marks)

d) In a third experiment a gas jar was filled to the brim with water. Salt was added one spatula measure at a time, until the water overflowed.

Twenty spatula measures were needed to reach this stage.

    (i) Why did the water not overflow when the first spatula measure of salt was added?

        (2 marks)

    (ii) Why did the water eventually overflow?

        (1 mark)

3 A motorist is inflating (blowing up) a car tyre with air.

a) Read the following statements carefully.

    A The inflated tyre is the same mass as the empty tyre.

    B The inflated tyre is heavier than the empty tyre.

    C The inflated tyre is lighter than the empty tyre.

Which statement **A**, **B** or **C** is correct?

        (1 mark)

b) Which special property of gases allows air to be used in the tyre?

        (1 mark)

c) After a long journey the motorist notices that the car tyres are warm.

What happens to the **volume** of air in the tyre as the tyres become warmer?

        (1 mark)

4   The diagrams show three **saturated** solutions of sugar in water.

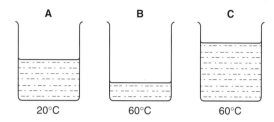

a) Which solution, **A, B** or **C**, contains the most sugar?

(*1 mark*)

b) Give **two** reasons why this solution contains the most sugar.

(*2 marks*)

c) Copy and complete the following sentence by choosing the correct word from the list below.

saturate   solute   solvent   solution

In a saturated solution of sugar in water, sugar is the _____.

(*1 mark*)

5   a) The solubility of a solid in water changes as the temperature changes.

What do you understand by the term 'solubility'?

b) A student investigated the solubility of two different solids at various temperatures. The table below shows the results obtained.

Temperature (°C)	Solubility of solid (g/100 g water)	
	Sodium chloride	Potassium chlorate
0	33.0	3.5
20	33.5	7.5
40	34.0	14.0
60	34.5	24.0
80	35.0	37.5

(i)   Plot the results for the two different solids on graph paper.

(*8 marks*)

(ii) Compare the effect of increasing the temperature on the solubility of **each** solid.

(*2 marks*)

(iii) At which temperature are the solubilities of the two solids the same?

(*1 mark*)

(iv) What is the solubility of the two solids at the temperature in part (iii)?

(*1 mark*)

(v)   If a saturated solution of potassium chlorate containing 50 g of water is cooled from 70°C to 30°C, what mass of the solid would crystallise?

(*4 marks*)

(vi) If a solution is made by dissolving 15 g of each solid in the same 100 g of water at 70°C and the solution is cooled to 20°C, crystals are observed. Which solid crystallises? Explain your answer with reference to both solids.

(*4 marks*)

6   If the volume and temperature of a fixed mass of gas are changed its new pressure can be calculated using the relationship

$$\frac{PV}{T} = \text{constant.}$$

Use this relationship and the information in the table below to calculate the unknown pressure. Show all steps in your calculation.

Conditions	Temperature (K)	Pressure (Pa)	Volume (cm³)
Initial	280	1400	300
Final	350		500

(*3 marks*)

Chapter 6

Chemical Equations

Learning objectives

By the end of this chapter you should be able to:

➤ Understand that what happens in a chemical reaction can be written as a word equation

➤ Write, and balance, equations with symbols/formulae instead of words and you will learn about state symbols

➤ Write equations that involve ions

Word and symbol equations

A chemical reaction can be written in the form of an **equation**. The substance or substances that react (the **reactants**) are on the left hand side and the substance or substances that are formed (the **products**) are on the right hand side.

You can use words or symbols/formulae:

Word equation:

  hydrochloric acid + sodium hydroxide = sodium chloride + water

Symbol equation:

$$HCl \quad + \quad NaOH \quad \rightarrow \quad NaCl \quad + \quad H_2O$$

It is important that you space the equation carefully especially when using words. If you cannot get all the words onto one line you must still keep the products on the right hand side of the equals sign or the arrow.

This is wrong:

sulphuric acid + potassium carbonate → potassium sulphate + carbon dioxide + water

This is better:

sulphuric acid + potassium carbonate → potassium sulphate + carbon dioxide + water

This is the best!

sulphuric acid + potassium carbonate → potassium sulphate + carbon dioxide + water

Questions

1 Write word equations for the following reactions:

a) Magnesium metal reacts with oxygen to form magnesium oxide.
b) When zinc carbonate is heated carbon dioxide gas is given off leaving zinc oxide.
c) Sulphur dioxide can be prepared by reacting hydrochloric acid with sodium sulphite. The other products are sodium chloride and water.

A word equation does not give as much information as a symbol equation because a symbol equation shows not only which chemicals are involved but also gives an indication of how much is used or produced.

sulphuric acid	+	sodium hydroxide	→	sodium sulphate	+	water
H_2SO_4	+	$2NaOH$	→	Na_2SO_4	+	H_2O
hydrochloric acid	+	sodium hydroxide	→	sodium chloride	+	water
HCl	+	$NaOH$	→	$NaCl$	+	H_2O

The symbol equation shows clearly that one sulphuric acid unit needs twice as much sodium hydroxide as does one hydrochloric acid unit.

Balancing equations

$$CaCO_3 \rightarrow CaO + CO_2$$

The above equation is said to be **balanced**. It has the same elements on both sides and the same number of atoms of each element.

$$NaHCO_3 \rightarrow Na_2CO_3 + CO_2 + H_2O$$

This equation is not balanced. It does have the same elements on both sides but not the same number of atoms of each element. Look at the sodium atoms; there is only one on the left hand side but there are two on the right hand side. The same is true of the hydrogen and carbon atoms. In the case of oxygen there are three atoms on the left hand side and six on the right hand side.

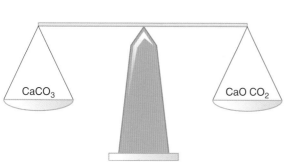

Figure 1 Both sides balance!

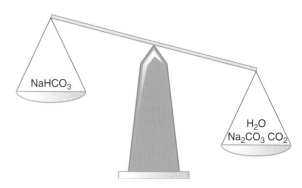

Figure 2 The two sides do not balance!

The **law of conservation of matter** is very important. It states that

> MATTER can be NEITHER CREATED nor DESTROYED

This law is applied in the writing of symbol equations.

Rules for writing symbol equations

- All the elements that appear on the left hand side are also found on the right hand side. (You cannot have an element in one of the reactants just disappearing nor can you make an element appear in the products that was not there to start.)
- Not only must you have the same elements present on both sides, you must also have the same number of atoms of each element.
- You cannot change the formula of a chemical to get the equation to balance.

Example

Reaction: sodium + water → sodium hydroxide + hydrogen

1 Write the correct formulae:

$$Na + H_2O \rightarrow NaOH + H_2$$

2 Check the Na: 1 on the left hand side (LHS) as the Na
1 on the right hand side (RHS) in the NaOH
so 1 on each side therefore the Na is balanced.

3 Check the H: 2 on the LHS in the H_2O
but 3 on the RHS, 1 in the NaOH and 2 in the H_2
to get more H on the LHS you must add another H_2O

(remember you cannot change the formula nor can you bring in any other chemicals so this is the only way to get more hydrogen)

$$Na + H_2O + H_2O \rightarrow NaOH + H_2$$

now there are 4 H on the LHS and only 3 on the RHS to get more hydrogen on the RHS add either another NaOH or a H_2,
an extra NaOH will give 4 H on each side, balanced.

$$Na + H_2O + H_2O \rightarrow NaOH + NaOH + H_2$$

4 Check the O: 2 on the LHS 1 in each of the 2 H_2O
2 on the RHS 1 in each of the 2 NaOH
so 2 on each side, balanced.

5 Recheck the Na: 1 on the LHS in the Na
2 on the RHS, 1 in each of the 2 NaOH
to get more Na on the LHS add another Na

$$Na + Na + H_2O + H_2O \rightarrow NaOH + NaOH + H_2$$

now there are 2 Na on each side, balanced.

It is not usual to write a formula more than once, numbers are put in front to show how many atoms or molecules are involved.

The balanced equation for the reaction between sodium and water is written as:

$$2Na + 2H_2O \rightarrow 2NaOH + H_2$$

It is not necessary to write out all of these stages, most of them are worked out in your head.

Example

$$sodium\ hydroxide\ +\ sulphuric\ acid\ \rightarrow\ sodium\ sulphate\ +\ water$$
$$NaOH\ +\ H_2SO_4\ \rightarrow\ Na_2SO_4\ +\ H_2O$$

Immediately it can be seen that the Na is not balanced, another NaOH is needed on the left hand side. You would not have to rewrite the equation but simply put a 2 in front of the NaOH

$$2NaOH + H_2SO_4 \rightarrow Na_2SO_4 + H_2O$$

Now check the other elements (done in your head, no need to write anything down unless you want to), put in any numbers that are needed for balance:

H   4 on the LHS in $2NaOH$ and H_2SO_4 and 2 on the RHS in the H_2O so another H_2O is needed, put a 2 in front of the H_2O

$$2NaOH + H_2SO_4 \rightarrow Na_2SO_4 + 2H_2O$$

S   1 on each side, in the H_2SO_4 and in the Na_2SO_4

O   6 on each side.

One short cut you could take would be to consider the SO_4 (or any other group that remains unchanged during the reaction) as single item.

Example

$$CaCl_2 + AgNO_3 \rightarrow AgCl + Ca(NO_3)_2$$

This equation has one Ca and one Ag on each side but the Cl and the NO_3 are not balanced. Putting a 2 in front of the AgCl balances the Cl but unbalances the Ag. A 2 has to be put in front of the $AgNO_3$ and this will not only balance the Ag but also the NO_3

$$CaCl_2 + 2AgNO_3 \rightarrow 2AgCl + Ca(NO_3)_2$$

Questions

2 Balance the following equations

a) $KHCO_3 + H_2SO_4 \quad \rightarrow \quad K_2SO_4 + CO_2 + H_2O$

b) $Mg + O_2 \quad \rightarrow \quad MgO$

c) $N_2 + H_2 \quad \rightarrow \quad NH_3$

d) $Fe_2O_3 + CO \quad \rightarrow \quad Fe + CO_2$

e) $C_4H_{10} + O_2 \quad \rightarrow \quad CO_2 + H_2O$

3 Write balanced symbol equations for the following word equations.

a) nitric acid + sodium hydroxide → sodium nitrate + water

b) hydrochloric acid + zinc carbonate → zinc chloride + carbon dioxide + water

c) silver nitrate + barium chloride → silver chloride + barium nitrate

d) magnesium sulphate + potassium hydroxide → magnesium hydroxide + potassium sulphate

e) Aluminium + sulphuric acid → aluminium sulphate + hydrogen

State symbols

At times it is important to know what state the reactants and products are in during the reaction.

The main **state symbols** are:

(s) **solid**   (l) **liquid**   (g) **gas**   (aq) **aqueous** (dissolved in water)

$$CaCO_3(s) + 2HCl(aq) \rightarrow CaCl_2(aq) + CO_2(g) + H_2O(l)$$

If a solvent other than water is used then the state symbol is the name of the solvent in brackets, for example NaOH (ethanol) represents sodium hydroxide dissolved in ethanol.

Ionic equations

There are times when an equation only needs to concentrate on some of the chemical species involved in the reaction.

Example: displacement of copper from its salt by magnesium.

$$CuSO_4 + Mg \rightarrow Cu + MgSO_4$$

This, however, does not really show what is happening to the copper and magnesium. If the sulphate, which is described as a **spectator ion**, is left out things become clearer.

$$Cu^{2+} + Mg \rightarrow Cu + Mg^{2+}$$

This **ionic equation** shows that the copper ion is becoming an atom and the magnesium atom is becoming an ion.

Half-equations, which include electrons and only deal with half of the reaction, show what is happening in these changes.

$$Cu^{2+} + 2e^- \rightarrow Cu$$

$$Mg \rightarrow Mg^{2+} + 2e^-$$

With ionic equations as well as balancing the elements and the number of atoms it is also necessary to balance the charges.

$$Cu^{2+} + Mg \rightarrow Cu + Mg^{2+}$$

Here the charge on each side is $2+$ and is balanced.

$$Cu^{2+} + 2e^- \rightarrow Cu$$

Here the charge on the left hand side is $2+$ plus $2-$ giving a total of 0 and that balances the 0 charge on the right hand side.

Questions

4   Balance the following ionic and half equations.

a) $Ag^+ + SO_4^{2-} \rightarrow Ag_2SO_4$
b) $Al^{3+} + e^- \rightarrow Al$
c) $Fe^{3+} + OH^- \rightarrow Fe(OH)_3$
d) $Cl^- \rightarrow Cl_2 + e^-$
e) $2H^+ + SO_4^{2-} + K^+ + OH^- \rightarrow K^+ + SO_4^{2-} + H_2O$

Websites

dbhs.wvusd.k12.ca.us/Equations/Equations.html

Exam questions

In examination papers questions asking you to write equations can appear within almost any topic.

For example:

- word or symbol equations for acid/base reactions,

- using state symbols to show precipitation,

- half equations in electrolysis.

Chapter 7

Oxidation and Reduction (Redox Reactions)

By the end of this chapter you will:

➤ Know what is meant by oxidation and reduction

➤ Recognise everyday examples of oxidation and reduction

➤ Know what happens with oxygen and hydrogen in oxidation and reduction

➤ Know what happens with electrons in oxidation and reduction

➤ Know how to work out what is being oxidised and what is being reduced

Figure 1 The burning of a match is a redox reaction

Figure 2 Food goes bad through oxidation

Some definitions

Oxidation and reduction in terms of oxygen

Oxidation is the **gain of oxygen** and so **reduction** which can be thought of as the opposite of oxidation, is the **loss of oxygen**.

Examples

$$2Mg + O_2 \rightarrow 2MgO$$

The magnesium has gained oxygen, it has been oxidised.

$$Ca + 2H_2O \rightarrow Ca(OH)_2 + H_2$$

The hydrogen has lost oxygen, it has been reduced.

Oxidation and reduction in terms of hydrogen

Oxidation is the **loss of hydrogen** and so **reduction** is the **gain of hydrogen**.

Examples

$$2HI \rightarrow H_2 + I_2$$

The iodine has been oxidised, it has lost hydrogen.

$$Cl_2 + H_2 \rightarrow 2HCl$$

The chlorine has gained hydrogen, it has been reduced.

Oxidation and reduction in terms of electrons

Oxidation is the loss of electrons and reduction is the gain of electrons.

Examples

$$2Cl^- \rightarrow Cl_2 + 2e^-$$

The chlorine has lost electrons, it has been oxidised. It has also changed from ions to a molecule.

$$Na^+ + e^- \rightarrow Na$$

The sodium has been reduced, it has gained electrons.

Displacement reactions are redox reactions. Consider the following:

$$Zn(s) + FeCl_2 (aq) \rightarrow ZnCl_2 (aq) + Fe(s)$$

Neither oxygen nor hydrogen is involved yet it is a redox reaction.

The zinc is changing from an atom to an ion and the iron is changing from an ion to an atom, there is a transfer of electrons.

$$Zn + Fe^{2+} \rightarrow Zn^{2+} + Fe$$

Half equations make the process clearer.

$$Zn \rightarrow Zn^{2+} + 2e^- \text{ loss of electrons = oxidation}$$

$$Fe^{2+} + 2e^- \rightarrow Fe \text{ gain of electrons = reduction}$$

How to remember oxidation and reduction in terms of oxygen and electrons

OXIDATION is SORE
Supply of
Oxygen
Removal of
Electrons

(Think of burning which involves reacting with oxygen, if you burn your finger it is sore.)

REDUCTION is ROSE
Removal of
Oxygen
Supply of
Electrons

Why don't you work out a method of remembering oxidation and reduction in terms of oxygen and hydrogen or even in terms of oxygen, hydrogen and electrons?

Where there is an oxidation there is always a reduction but it is not always easy to see both processes.

When hydrogen reacts with copper oxide

$$H_2 + CuO \rightarrow H_2O + Cu$$

It is easy to see that the hydrogen is oxidised (it gains oxygen) and the copper is reduced (it loses oxygen).

Look back at the reaction between magnesium and oxygen:

$$2Mg + O_2 \rightarrow 2MgO$$

The oxidation of magnesium is easy to see but where is the reduction? It must be happening to the oxygen, as it is the only other element present. Look carefully at the oxygen, it is a molecule with two oxygen atoms but there is only one oxygen ion in each magnesium oxide. You could explain the reduction of oxygen by saying that each oxygen atom loses the other one.

Another way to explain the reduction is to look at the change in the oxygen from two atoms in a molecule to two oxygen ions:

$$O_2 + 4e^- \rightarrow 2O^{2-}$$

The oxygen is gaining electrons.

Look back at the reaction between chlorine and hydrogen.

$$Cl_2 + H_2 \rightarrow 2HCl$$

The gain of hydrogen by the chlorine and therefore its reduction is easy to see, the oxidation is not so clear. The only other element present is hydrogen and so it must be oxidised. Like oxygen in the previous reaction there are two hydrogen atoms in the molecule but only one hydrogen atom in the molecule of hydrogen chloride. Each hydrogen can be said to lose the other one and so the hydrogen is oxidised.

Again the explanation can be given in terms of electrons. If the hydrogen chloride were to be ionised then hydrogen ion would be formed.

$$H_2 \rightarrow 2H^+ + 2e^-$$

The hydrogen is losing electrons

Oxidising and reducing agents

An **oxidising agent** 'does the oxidising' that is it supplies the oxygen, removes the hydrogen or removes the electrons . It follows then that the oxidising agent must itself be reduced, it loses oxygen, gains hydrogen or gains electrons .

The opposite is true for the **reducing agent**.

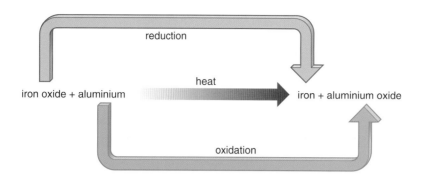

Figure 3 The redox reaction between iron oxide and aluminium

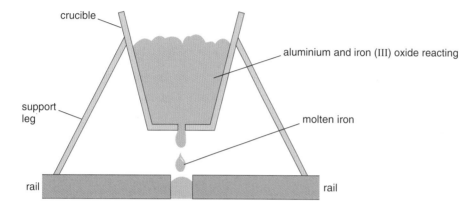

Figure 4 The Thermit process for welding together two lengths of railway line

Consider the reaction between iron oxide and aluminium, a very exothermic reaction that is used to weld railway lines together.

$$Fe_2O_3 + 2Al \rightarrow 2Fe + Al_2O_3$$

The iron oxide supplies the oxygen that oxidises the aluminium so the iron oxide is the oxidising agent. At the same time the aluminium is removing the oxygen from the iron so the aluminium is the reducing agent.

In terms of electrons:

$Fe^{3+} + 3e^- \rightarrow Fe$ gain of electrons = reduction

$Al \rightarrow Al^{3+} + 3e^-$ loss of electrons = oxidation

Figure 5 Divers use thermit to weld underwater

Questions

1   For each of the following name:
   (i)   the element being oxidised
   (ii)   the element being reduced
   (iii) the oxidising agent
   (iv)   the reducing agent.

   a) $ZnO + C \rightarrow Zn + CO$
   b) $Mg + H_2O \rightarrow MgO + H_2$
   c) $CuS + O_2 \rightarrow Cu + SO_2$
   d) $Mg + CuSO_4 \rightarrow Mg\,SO_4 + Cu$
   e) $Cl_2 + 2NaBr \rightarrow 2NaCl + Br_2$

Important redox reactions

Rusting: when iron rusts it forms an oxide by gaining oxygen, the iron is oxidised. Water is also needed.

$$2Fe + 3O_2 + 2H_2O \rightarrow Fe_2O_3 . 2H_2O$$

Combustion of fuels: combustion is the reaction with oxygen forming an oxide and releasing energy. This means that when a fuel, for example methane, burns an oxidation is taking place.

$$CH_4 + 2O_2 \rightarrow CO_2 + 2H_2O$$

Both the carbon and the hydrogen have gained oxygen.

Extraction of aluminium and of iron: aluminium and iron are found naturally as their oxides. The oxygen is removed to obtain the metal so the aluminium or iron loses oxygen and is reduced.

$$2Al_2O_3 \rightarrow 4Al + 3O_2 \text{ by electrolysis}$$

$$Fe_2O_3 + 3CO \rightarrow 2Fe + 3CO_2 \text{ in the blast furnace}$$

Manufacture of ammonia: ammonia is formed by the reaction of nitrogen with hydrogen; the nitrogen is gaining hydrogen and is reduced.

$$N_2 + 3H_2 \rightarrow 2NH_3$$

Respiration: the breakdown of glucose in the cells of living organisms is a special example of combustion. Oxygen is required and oxides are formed.

$$C_6H_{12}O_6 + 6O_2 \rightarrow 6CO_2 + 6H_2O$$

The carbon in the glucose is gaining oxygen and is being oxidised.

Manufacture of sulphuric acid: the main reaction involves sulphur dioxide reacting with oxygen to form sulphur trioxide, as the sulphur is gaining more oxygen it is being oxidised.

$$2SO_2 + O_2 \rightarrow 2SO_3$$

Websites

www.gcsechemistry.com/ukop/htm

Exam questions

1   This question is about the reaction that happens when burning magnesium is lowered into a jar of carbon dioxide.

   a) Give a word equation for the reaction.
   (1 mark)

   b) Explain fully why the reaction between magnesium and carbon dioxide is described as a **redox** reaction.
   (5 marks)

2   This question is about the reaction that happens when magnesium metal is added to copper sulphate solution.

   a) Write a balanced symbol equation for the reaction.
   (2 marks)

   b) The reaction of magnesium with copper sulphate may be described as a redox reaction. Explain what this term means when applied to this reaction.
   (5 marks)

Chapter 8

Acids, Bases and Salts

Learning objectives

By the end of this chapter you will:

➤ Know how to recognise acids and bases

➤ Have learnt about typical reactions of acids and how to name the salts that are formed

➤ Know about different methods of preparing salts

➤ Know about indicators and the pH scale for measuring acidity

➤ Know how to test for the sulphate, carbonate and halide anions

Acids and alkalis in everyday life

There are many acids in everyday life. Examples are ethanoic acid in vinegar, citric acid in lemon juice and carbonic acid in soda water. Acids taste sour but tasting them would not be recommended as a test for acidity! Examples of alkalis can also be found in the home. Many alkalis feel soapy but again this is not a suitable test.

Some definitions

Acids

Acids contain hydrogen and when they dissolve in water they produce hydrogen ions

$$HCl \ + \ aq \ \rightarrow \ H^+(aq) \ + \ Cl^-(aq)$$

$$H_2SO_4 \ + \ aq \ \rightarrow \ 2H^+(aq) \ + \ SO_4{}^{2-}(aq)$$

Acids react to form salts and the acid determines the type of salt formed.

ACID	SALT
hydrochloric	chloride
nitric	nitrate
sulphuric	sulphate

Bases

A base reacts with an acid to form a salt and water.

$$\boxed{\text{acid} + \text{base} \rightarrow \text{salt} + \text{water}}$$

Metal oxides and hydroxides are bases.

The base determines exactly which salt is formed

copper oxide + *sulphuric* acid → **copper** *sulphate* + water

magnesium oxide + *nitric* acid → **magnesium** *nitrate* + water

An **alkali** is a soluble base and so

$$\boxed{\text{acid} + \text{alkali} \rightarrow \text{salt} + \text{water}}$$

The relationship between bases and alkalis is shown in the Venn diagram (Figure 1).

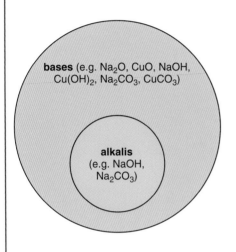

Figure 1 The relationship between alkalis and bases

When an alkali dissolves in water it produces, directly or indirectly, hydroxide ions.

$$NaOH + aq \rightarrow Na^+(aq) + OH^-(aq)$$

$$NH_3 + H_2O \rightarrow NH_4^+(aq) + OH^-(aq)$$

Neutralisation

When an acid and an alkali react the hydrogen and hydroxide ions form water. Water is neutral and so this type of reaction is called a **neutralisation**.

$$H^+(aq) + OH^-(aq) \rightarrow H_2O(l)$$

Example   sulphuric acid + sodium hydroxide → sodium sulphate + *water*

Other reactions of acids

Acids react with metals

metal + acid → metal salt + hydrogen

Example $Zn + H_2SO_4 \rightarrow ZnSO_4 + H_2$

Acids react with carbonates

acid + carbonate → salt + carbon dioxide + water

Example $2HCl + MgCO_3 \rightarrow MgCl_2 + CO_2 + H_2O$

Questions

1 Complete the following equations.

a) copper hydroxide + hydrochloric acid → ———— + water

b) sodium hydroxide + sulphuric acid → ———— + water

c) calcium carbonate + nitric acid → ———— + carbon dioxide + water

d) magnesium + ———— acid → magnesium chloride + hydrogen

e) ———— + sulphuric acid → sodium sulphate + carbon dioxide + water

f) ———— oxide + ———— acid → zinc nitrate + water

g) ———— hydroxide + ———— acid → potassium chloride + water

Indicators

It is often useful to know if a solution is **acidic**, **neutral** or **alkaline**. **Indicators** are substances that change to a particular colour depending on which type of solution they are in. Litmus is one of the best-known indicators; it can be used as a solution or as litmus paper.

INDICATOR	Colour in acidic soln	Colour in neutral soln	Colour in alkaline soln
Litmus	red	purple	blue
Methyl orange	red/pink	yellow	yellow
Phenolphthalein	colourless	colourless	pink

Table 1 The colours of indicators

Indicators are made from coloured material extracted from plants. Making your own indicator is fairly simple. You need brightly coloured juice from a plant, beetroot and raspberries are good. If the plant does not produce a juice easily try grinding it with some hot water. To check the usefulness of the juice as an indicator add a drop of it to some acid and another drop to some alkali. A definite difference in colour means that the juice could be used as an indicator. To try this at home use vinegar as the acid and a solution of washing soda as the alkali.

Questions

2 IT: Use some Universal Indicator paper (pH paper) to test as many household substances as you can. Put your results on a spreadsheet and organise them into two lists, one in alphabetical order and one in order of increasing pH. Write a report of your findings using a suitable word processing package.

pH Scale

You must be careful how you write **pH**. Even at the beginning of a sentence the p is small. The H is always a capital because it is the symbol for hydrogen.

The **pH scale** is a measure of how acidic or alkaline a solution is. pH is measured by a pH meter or by using Universal Indicator paper, which goes different colours for different pH values.

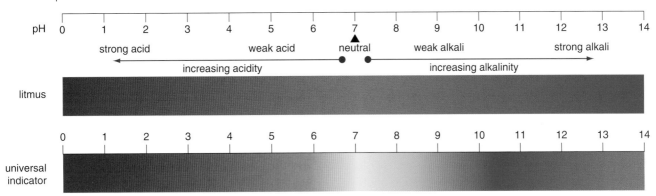

Figure 2 The colours of litmus and of universal indicator in solutions of different pH

A **neutral** solution has a pH of 7, below 7 a solution is acidic and above 7 a solution is alkaline. The lower the pH value the stronger the acid and the higher the pH the stronger the alkali. How the colours of litmus and of universal indicator change with pH is shown in Figure 2.

Salt preparation

The method used depends on the solubility of the salt.

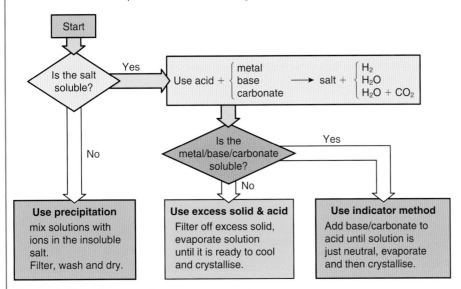

Figure 3 Flow diagram: how to prepare salt

A soluble salt from an acid and an alkali in solution

Example: sodium chloride from dilute hydrochloric acid and aqueous sodium hydroxide

$$HCl + NaOH \rightarrow NaCl + H_2O$$

> ### Did you know?
>
> Hydrochloric acid produced in your stomach gives the contents a pH of around 2.
>
> The acidic mixture being pushed out of your stomach and part way up the oesophagus causes heartburn.
>
> Your blood, however, has a pH of 7.4.

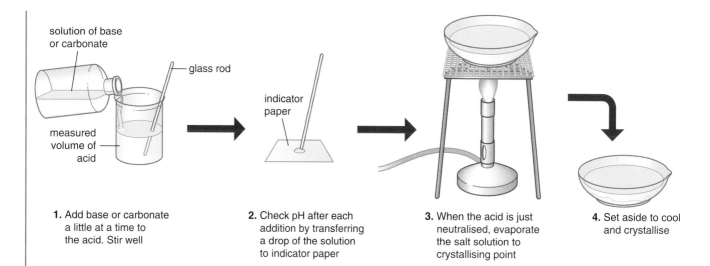

1. Add base or carbonate a little at a time to the acid. Stir well

2. Check pH after each addition by transferring a drop of the solution to indicator paper

3. When the acid is just neutralised, evaporate the salt solution to crystallising point

4. Set aside to cool and crystallise

Figure 4 How to prepare a soluble salt from a soluble base or carbonate

This involves adding one colourless liquid to another to make a third. As a result it is impossible to see when to stop adding the second solution. An indicator can be used to determine when the mixture is neutral.

Safety note: if the mixture is not neutral when it is heated then the salt obtained will not be pure. If the acid is in excess then a concentrated acid is produced which is undesirable. Concentrated alkali solution and solid alkali can also cause problems.

A soluble salt from an acid and an insoluble base

Example: copper sulphate from sulphuric acid and copper oxide

$$H_2SO_4 + CuO \rightarrow CuSO_4 + H_2O$$

The solid copper oxide is added to a set volume of the acid until there is excess oxide in the mixture. This is to make sure that all the acid has reacted and it will not be a problem when heating to obtain the salt. Filtration is needed to remove the excess oxide and then the filtrate is heated to obtain the salt.

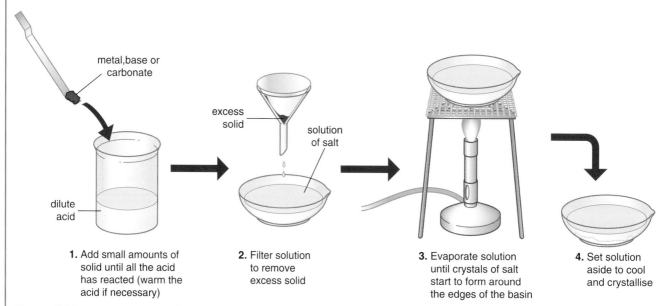

1. Add small amounts of solid until all the acid has reacted (warm the acid if necessary)

2. Filter solution to remove excess solid

3. Evaporate solution until crystals of salt start to form around the edges of the basin

4. Set solution aside to cool and crystallise

Figure 5 How to prepare a soluble salt from a metal, or from an insoluble base or carbonate

Because copper sulphate can exist in the hydrated and in the anhydrous form the degree of heating depends on which is required. The anhydrous form is easily obtained by heating the filtrate to dryness. If the hydrated form is required, however, the mixture is only heated until the first ring of solid appears around the edge of the evaporating basin. This gives a hot concentrated solution that is then left to cool and crystals of the hydrated salt form. These crystals are removed from the mixture by filtration or decanting off the liquid. Finally they are patted dry with paper towels or filter paper.

An insoluble salt by precipitation

A **precipitate** is a suspension of fine solid particles in a liquid, formed during a chemical reaction. The precipitate is insoluble and as it forms it comes out of the solution as a solid.

Insoluble salts are prepared by mixing two solutions, one of which contains the required cation and the second contains the anion. The reaction involves the ions 'changing partners'.

sodium carbonate + *zinc sulphate* → **sodium** *sulphate* + *zinc* **carbonate**

In this example the precipitate is zinc carbonate.

Example: preparation of lead bromide

Lead nitrate solution reacts with sodium bromide solution to give a precipitate of lead bromide and a solution of sodium nitrate.

$$Pb(NO_3)_2(aq) + 2NaBr(aq) \rightarrow PbBr_2(s) + 2NaNO_3(aq)$$

The mixture is filtered and the lead bromide is left in the filter paper. It is washed with distilled water to remove any sodium nitrate solution and then is left to dry.

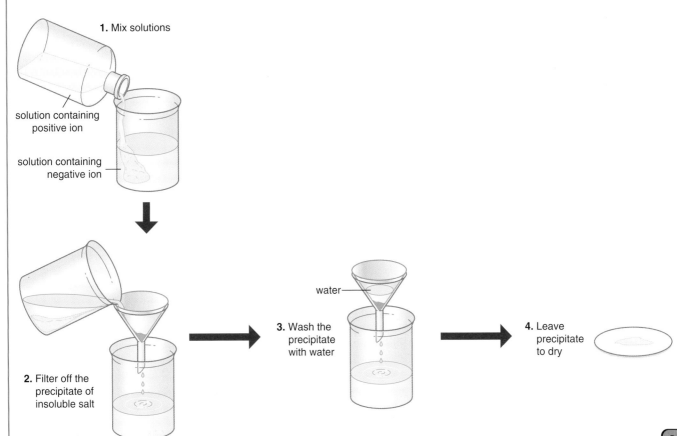

1. Mix solutions

solution containing positive ion

solution containing negative ion

2. Filter off the precipitate of insoluble salt

3. Wash the precipitate with water

water

4. Leave precipitate to dry

Figure 6 Preparing an insoluble salt

Questions

3   Complete the following and underline the precipitate in each one.

a)   barium   + potassium   → barium   + _____
     nitrate   sulphate   sulphate   _____

b)   sodium   + lead   → sodium   + _____
     carbonate   nitrate   nitrate   _____

c)   magnesium + sodium   → magnesium + _____
     chloride   hydroxide   hydroxide   _____

4   Give the name of the acid needed to make each of the following salts and
    suggest a suitable chemical that would react with the acid to form the salt.

a)   Potassium sulphate
b)   Magnesium chloride
c)   Zinc nitrate.

5   Which of the following statements are true?

a)   Lead nitrate can be prepared by precipitation
b)   Copper carbonate can be prepared by reacting copper sulphate solution
     with sodium carbonate solution
c)   Magnesium hydroxide would react with nitric acid to give magnesium
     nitrate and water
d)   A solution of iron(II) carbonate can be reacted with sulphuric acid to
     form iron(II) sulphate
e)   A salt solution must not be heated to dryness if a sample of the
     hydrated salt is wanted.

Tests for Anions

Carbonate

To a small amount of the solid sample add dilute hydrochloric acid. If the
sample is a carbonate carbon dioxide will be given off. This is best done in a
test tube so that the gas can be collected in a dropper and bubbled through
limewater.

$$CaCO_3 + 2HCl \rightarrow CaCl_2 + CO_2 + H_2O$$

Halide (chloride, bromide or iodide)

Dissolve a small amount of the solid sample in distilled water. Add about the
same volume of dilute hydrochloric acid followed by 2–3 drops of silver nitrate
solution. If you see a precipitate (of the silver halide) then the sample is a
halide. Chlorides give a white precipitate, bromides a cream one and iodides a
yellow one.

$$NaCl(aq) + AgNO_3(aq) \rightarrow NaNO_3(aq) + AgCl(s)$$

Unfortunately the differences in these colours are not always clear. Adding dilute ammonia solution helps to distinguish between them. Silver chloride will dissolve, silver bromide will only dissolve with a large excess of ammonia (or by using concentrated ammonia solution) but the silver iodide will not dissolve at all.

Sulphate

Dissolve a small amount of the solid sample in distilled water. Add about the same volume of dilute hydrochloric acid followed by 2–3 drops of barium chloride solution. A white precipitate of barium sulphate means that the sample was a sulphate.

$$MgSO_4(aq) + BaCl_2(aq) \rightarrow MgCl_2(aq) + BaSO_4(s)$$

Limits to classification

In chemistry we have many ways of grouping, or classifying, chemicals or reactions.

Examples: solid, liquid and gas,
element, compound and mixture
metal and non-metal

While dividing substances into these and other groups can be very useful it is important to realise that there can be exceptions.

Oxides

As a general rule metal oxides are basic and non-metal oxides are acidic.

Typical metal oxide:

ACID	+	BASE	→	SALT	+ WATER
hydrochloric acid	+	copper oxide	→	copper sulphate	+ water

Typical non-metal oxide:

ACID	+	BASE	→	SALT	+ WATER
carbon dioxide		sodium hydroxide	→	sodium carbonate	+ water

Exceptions to this rule include water (hydrogen oxide) and carbon monoxide that are neither acidic nor basic but are neutral.

Acids and bases

There are certain chemicals that can behave either as an acid or a base; they are described as being amphoteric. Both aluminium and zinc hydroxides are amphoteric.

ACID	+	BASE	→	SALT	+	WATER
hydrochloric acid	+	**aluminium hydroxide**	→	aluminium chloride	+	water
ACID	+	BASE	→	SALT	+	WATER
aluminium hydroxide	+	sodium hydroxide		sodium aluminate	+	water
ACID	+	**BASE**	→	SALT	+	WATER
sulphuric acid	+	**zinc hydroxide**	→	zinc sulphate	+	water
ACID	+	BASE	→	SALT	+	WATER
zinc hydroxide	+	sodium hydroxide		sodium zincate		water

Websites

www.chem.ubc.ca/courseware/pH/launch.html

www.gcsechemistry.com/ukop.htm

www.s-cool co.uk

Exam questions

1 a) Cartons of fruit juice contain some acid.

(i) Describe how you could show that the apple juice is slightly acidic.
(2 marks)

(ii) What pH would you expect slightly acidic apple juice to have?
(1 mark)

(iii) What ion is always present in an acid solution?
(1 mark)

b) Oxides may be classified as acidic, basic or neutral.

Copy and complete the table below of oxide classifications. The first one has been done for you.

Oxide	Type
Nitrogen dioxide	acidic
Water	
Sulpher dioxide	
Copper(II) oxide	
Carbon dioxide	
Magnesium oxide	
Calcium oxide	

(3 marks)

2 This question is about acids, bases and salts.

The table below gives the pH values of a number of solutions of household substances.

Substance	pH value
Baking soda	9
Battery acid	1
Milk	6
Orange juice	4
Oven cleaner	13
Vinegar	3
Washing soda	11
Water	7

a) Why can litmus **not** be used to measure pH values? *(1 mark)*

b) Which substances, listed in the table above, is the strongest alkali? *(1 mark)*

c) Ant stings contain a weak acid. From the table above name **two** substances which could be used to neutralise the acid. *(2 marks)*

d) Hydrochloric acid is strong acid.
   (i) What pH value would you expect for hydrochloric acid? *(1 mark)*
   (ii) What is the chemical formula for hydrochloric acid? *(1 mark)*

e) Acids react with bases and metals to form salts.
   (i) What name is given to a soluble base? *(1 mark)*
   (ii) Copy and complete the word equations below.

[A]

$$\underline{\hspace{1.5cm}} + \text{sodium hydroxide} \rightarrow \text{sodium chloride} + \text{water}$$

[B]

$$\text{hydrochloric acid} + \underline{\hspace{1.5cm}} \rightarrow \text{magnesium chloride} + \text{hydrogen}$$

[C]

$$\text{sulphuric acid} + \text{copper (II) oxide} \rightarrow \underline{\hspace{1.5cm}} + \underline{\hspace{1.5cm}}$$

(4 marks)

   (iii) A student wanted to make sodium sulphate by reacting sodium metal with dilute sulphuric acid. Why should this method not be used? *(2 marks)*
   (iv) Salts are ionic solids. Give **two** typical properties of salts. *(2 marks)*

3 Methyl orange and phenolphthalein are two useful indicators in the laboratory. They change colour when they are mixed with acids or alkalis.

Indicator	Colour in acid	Colour in neutral solution	Colour in alkali
Methyl orange	pink	yellow	yellow
Phenolphthalein	colourless	colourless	red

a) Using the table above, copy and complete the following table.

Aqueous solution	Colour of solution in the presence of:	
	Methyl orange	Phenolphthalein
Hydrogen chloride		
Ammonia		
Sodium chloride		
Sulpher dioxide		

(4 marks)

b) A solution is made up by mixing 25 cm³ of NaOH(aq) with 15 cm³ of H_2SO_4(aq). Three drops of methyl orange indicator are added.

The NaOH(aq) and H_2SO_4(aq) were of equa concentration before addition. What colour would you expect for the resultant solution. *(1 mark)*

c) Would the indicators above be suitable to decide if a solution of bromine in water is acidic, neutral or alkaline? Explain your answer. *(3 marks)*

d) Copper nitrate, a blue crystalline solid, may be obtained by reacting copper carbonate with dilute nitric acid and crystallising the resulting solution. The following method is often followed:

50 cm³ of nitric acid, concentration 1 mol/dm³ (moles per litre) are placed in a 250 cm³ beaker and copper carbonate is added with stirring until no further reaction takes place. The mixture is filtered into an evaporating basin and the solution reduced to about $\frac{1}{3}$ in volume by evaporation. The liquid is left to cool until blue crystals have formed. The crystals are filtered off, washed with a little cold water and dried between filter papers. Finally, the crystals are weighed.

(i) Write a balanced, symbol equation for the reaction. *(2 marks)*

(ii) How would you know when the reaction had finished? *(1 mark)*

(iii) What was the purpose of the first filtration? *(2 marks)*

(iv) Why was the final filtrate not evaporated to dryness? *(1 mark)*

(v) In an experiment to make copper nitrate a student obtained 5.50 g of crystals. Calculations show that 6.05 g of crystals could have been obtained from the amounts of the starting materials which were used. Suggest **two** reasons why the student did not obtain the full 6.05 g of crystals. *(2 marks)*

4 a) Zinc sulphate, a white crystalline solid, may be obtained by reacting zinc metal with dilute sulphuric acid and crystallising the resulting solution.

50 cm³ of dilute sulphuric acid were placed in a beaker and zinc granules were added. When the reaction had finished, the mixture was filtered into an evaporating basin. The solution was reduced to about $\frac{1}{3}$ volume and then left to cool until white crystals had formed. The crystals were filtered off, washed with a little cold water and dried between filter papers.

(i) Write a **symbol** equation for the reaction. *(2 marks)*

(ii) Describe what would be **observed** when the zinc is added to the acid. *(2 marks)*

(iii) How would you know when the reaction had finished? *(1 mark)*

(iv) What is the residue from the first filtration? *(2 marks)*

(v) Why is the solution reduced to about $\frac{1}{3}$ volume? *(1 mark)*

(vi) Why are the crystals washed with water? *(1 mark)*

b) Zinc oxide can also be used to make zinc sulphate crystals by a similar method.

(i) How would you know when this reaction had finished? *(1 mark)*

(ii) Give a **symbol** equation for the reaction of zinc oxide with dilute sulphuric acid. *(2 marks)*

c) The method of salt preparation used in part a) cannot be used with either sodium or copper metal. In each case give a reason why this method cannot be used. *(2 marks)*

d) Zinc sulphate is a soluble salt. Some salts such as barium sulphate and lead chloride are insoluble. By which method are insoluble salts made? *(1 mark)*

6 An understanding of the chemistry of acids and alkalis is important since many substances which we use every day are either acidic or alkaline.

a) Universal indicator is commonly used in the laboratory. Complete the table below.

Indicator	Colour in sulphuric acid	Colour in sodium hydroxide
Universal indicator		

(2 marks)

b) Acids react with **bases** and **alkalis** to form salts. Explain carefully what is meant by the terms. *(3 marks)*

c) Explain carefully how you would make **pure dry** crystals of copper sulphate starting from copper oxide. Include all observations and write a **symbol** equation for the reaction which occurs. *(10 marks)*

d) An important skill in Chemistry is to be able to identify compounds using various analytical tests. Describe how you could test the salt prepared in part c) about to prove that it was in fact copper sulphate.
(i) Test for copper ions. *(4 marks)*
(ii) Test for sulphate ions. *(4 marks)*

Chapter 9

Water

By the end of this chapter you will:

➤ Know about water, the chemical, and how to test for its presence

➤ Know about water of crystallisation

➤ Know about hard water: how to identify it, its causes and types and how to make it soft. You will also learn about the problems it creates

➤ Know which ions cause hard water and how they behave when the water is softened

➤ Know how water can be polluted and how it is treated to make it fit for human consumption

Some characteristics of water

Water is the commonest compound on Earth and it is vital for life. It is colourless, odourless, and an excellent solvent for many substances. It is neutral, it does not burn nor does it allow burning.

Pure water is not found naturally because it is such a good solvent; even seemingly very clean water will contain some solutes. Oxygen and nitrogen dissolve in water, to a very small extent certainly, but they will be present in any sample.

It has a melting point of $0°C$, a boiling point of $100°C$ and a density of 1 g/cm^3.

Water of crystallisation and the test for water

Molecules of water are found with some salts in their **hydrated** form. These water molecules form part of the crystal structure and that is why they are called **water of crystallisation**.

Heating can decompose the crystals and the water molecules are separated from the salt leaving the **anhydrous** form of the salt.

In several examples the colour of the hydrated crystals differs from that of the anhydrous powder.

$$CuSO_4.5H_2O \quad \rightarrow \quad CuSO_4 \quad + \quad 5H_2O$$
$$\text{blue} \qquad\qquad \text{white}$$

$$CoCl_2.2H_2O \quad \rightarrow \quad CoCl_2 \quad + \quad 2H_2O$$
$$\text{pink} \qquad\qquad \text{blue}$$

These reactions can be reversed and can be used to test for the presence of water. A few drops of the liquid being tested are placed on the anhydrous powder and if the given colour change happens then water must be present in the liquid sample.

$$CuSO_4 \quad + \quad 5H_2O \quad \rightarrow \quad CuSO_4.5H_2O$$

white   blue

$$CoCl_2 \quad + \quad 2H_2O \quad \rightarrow \quad CoCl_2.2H_2O$$

blue   pink

The test only shows that water is present and does not give any information about the purity.

Cobalt chloride paper is made by soaking paper in a solution of cobalt chloride and then drying it to leave the paper impregnated with the anhydrous salt. A strip of this paper can be dipped into a liquid to test for the presence of water which then turns the paper blue to pink. Unfortunately the paper will absorb water from the air and turn from blue to pink. This means that it should be kept somewhere dry or else it has to be dried before being used.

Interaction of substances with moist air

Think of common salt left in a container in damp weather, it becomes damp and sticky. The salt has taken in water from the air. Salt (sodium chloride) is not the only substance to do this. Remember sodium has to be kept under oil because it would react with the moisture in the air.

This taking in of water from the air can also be useful. Silica gel makes a good **desiccant** (drying agent) because it absorbs water from the air. That is why little sachets of silica gel are placed with many items when they are packed for storage or transport, for example optical instruments.

Anhydrous calcium chloride is another useful drying agent especially in the laboratory. It is cheap because it is a waste product in the manufacture of sodium carbonate.

If sodium hydroxide pellets are left exposed to moist air they can take in so much water that they dissolve in it.

Hard water

Water that does not lather easily with soap is called **hard water**. If the water does lather easily it is **soft water**.

When soap is used with hard water not only is it difficult to get a good foamy lather but there is also a **scum** formed.

The soap molecules are reacting with the soap to form the scum.

hard water + soap → scum

Hard water will produce lather eventually but it uses more soap than a sample of soft water of the same volume.

Soapless detergents will lather well with both hard and soft water because it is only soap that reacts to form scum.

Figure 1 Scum in hard water

Causes of hard water

Solution	Behaviour with soap solution
Sodium chloride	Good lather
Calcium chloride	No lather, scum
Magnesium chloride	Poor lather, scum
Potassium nitrate	Good lather
Calcium nitrate	Poor lather, scum
Magnesium nitrate	No lather, scum
Magnesium sulphate	No lather, scum
Sodium sulphate	Good lather

Table 1

From the table it can be seen that dissolved calcium or magnesium ions make water hard. The calcium or magnesium ions react with soap forming insoluble salts which are precipitated out of the mixture as the scum.

$$Ca^{2+}(aq) + 2C_{17}H_{35}COONa(aq) \rightarrow (C_{17}H_{35}COO)_2Ca(s) + Na^+(aq)$$
sodium stearate (soap)   calcium stearate (scum)

Types of hard water

Temporary hard water

This can be softened by boiling.

Carbon dioxide in the air dissolves in rain water forming the weak acid, carbonic acid.

$$CO_2(g) + H_2O(l) \rightarrow H_2CO_3(aq)$$

When this dilute acidic solution falls on rocks containing calcium carbonate e.g. limestone, soluble calcium hydrogencarbonate is formed. As a result there are calcium ions dissolved in water and the water is hard.

$$H_2CO_3(aq) + CaCO_3(s) \rightarrow Ca(HCO_3)_2(aq)$$

The calcium hydrogencarbonate, however, is not very stable and is easily decomposed on heating. The calcium carbonate produced is insoluble and is precipitated out of the solution.

$$Ca(HCO_3)_2(aq) \rightarrow CaCO_3(s) + CO_2(g) + H_2O(l)$$

As the calcium ions are removed the hard water is softened. Unfortunately the calcium carbonate is deposited in the heating vessel.

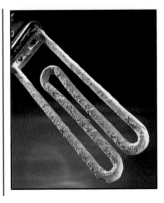

Figure 2 'Fur' inside a kettle is a deposit of calcium carbonate from hard water

Permanent hard water

Cannot be softened by boiling.

Carbonates are not the only calcium- or magnesium-containing compounds found in rocks, gypsum, for example, contains calcium sulphate. When rain passes over or through these other rocks, calcium or magnesium ions can dissolve in the water and make it hard. Because these compounds do not decompose on heating the calcium or magnesium ions remain in the water and it stays hard.

$$CaSO_4(s) + aq \rightarrow Ca^{2+}(aq) + SO_4^{2-}(aq)$$

Question

1 $20\,cm^3$ was taken from each of 5 samples of water and they were shaken with soap solution. The volume of soap solution needed to get a good lather was noted.

The samples were then boiled and the test was repeated.

Sample		A	B	C	D	E
vol. of soap sol. (cm^3)	before boiling	10.1	6.3	12.2	1.6	8.4
	after boiling	10.1	1.9	2.3	1.6	8.4

a) Which letter represents the sample of soft water?
b) Which letter represents the sample of the hardest water?
c) Which letters represent the two samples of temporary hard water?

Softening hard water

The three main methods are boiling, adding washing soda and using an ion exchanger.

● **Boiling** can be used but only for temporary hard water.

● **Adding washing soda**, hydrated sodium carbonate $Na_2CO_3.10H_2O$, which precipitates out the calcium as its carbonate.

$$Na_2CO_3(aq) + Ca(HCO_3)_2(aq) \rightarrow 2NaHCO_3(aq) + CaCO_3(s)$$

● **Using an ion exchanger** that removes the calcium or magnesium ions and replaces them with other suitable ions.

Advantages and disadvantages of hard water

Advantages of hard water

● Calcium is needed for strong bones and teeth and may help to prevent heart disease. You would, however, have to drink a large amount of water in order to get a significant amount of calcium!

● Hard water is supposed to taste better but this is presumably a matter of individual taste. Brewers find hard water is better for making beer.

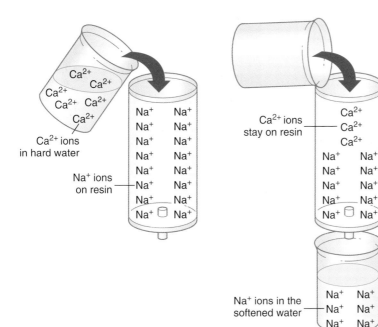

Figure 3 Using an ion exchange column

- Hard water is less likely to dissolve heavy metals such as lead. So if the hard water is flowing through old lead pipes then there is a smaller possibility of the lead dissolving in the water and damaging your health.

- In tanning leather the calcium or magnesium ions make the leather cure better.

Disadvantages of hard water

- Hard water will, of course, produce a lather with soap eventually, but the amount of soap required makes it much more expensive than using soft water.

- When temporary hard water is heated in a kettle the calcium carbonate deposited builds up on the inside and, if it is an electric kettle, on the element. This is known as 'fur'. Further use of the kettle needs more heat as the 'fur' has to be heated before the energy gets to the water.

- When the hot water is in boilers and pipes the deposits are known as 'scale' or 'limescale'. Again this causes problems. The pipes can become blocked and the hot water system is much less efficient.

Characteristics of a hard water area

Hard water comes from areas with chalk or limestone. As the rain, containing the carbonic acid, runs through these rocks some of the calcium carbonate is removed as the soluble calcium hydrogencarbonate.

Figure 4 Carbonic acid in rain water has helped to diissolve the limestone in the cracks of this limestone pavement

Over very many centuries this can create caves. Inside the caves **stalagmites** and **stalactites** can form. A drop of temporary hard water on the roof or floor of the cave can lose water by evaporation. As a result the calcium hydrogencarbonate decomposes leaving a solid deposit of calcium carbonate. (This is the same reaction that takes place in the softening of temporary hard water by boiling.) Very, very slowly the deposits build up to form a column of calcium carbonate.

Figure 5 Stalagmites and stalactites

Treatment of water

Water for our homes, schools, factories and other buildings comes from many different sources such as lakes, rivers and reservoirs. Taken straight from any of these sources it may be suitable for several uses but it is unlikely to be fit for humans to drink.

The treatment of water can involve a variety of processes, depending on the origin of the water, but there are two main treatments.

> **Did you know?**
>
> Stalagmites rise up from the floor of the cave and stalactites hang down from the ceiling. One way to remember is:
>
> stalag**M**ites **M**ount up
>
> stalac**TITE**s cling tightly (*TITE*ly) to the ceiling.

1 Filtering. The water is allowed to run through sand beds to remove insoluble solid matter. There may be two different sand beds: first coarse sand to remove larger solids and after that fine sand to remove smaller particles.

2 Chlorination. Microbes in water cause several diseases, e.g. cholera, typhoid and other lesser problems. Chlorine is added to water in order to kill the bacteria. Great care is needed in order to use enough chlorine to kill the micro-organisms but not enough to kill or harm anything bigger.

Because chlorine in water gives an acidic solution the pH of the water may have to be adjusted before the water leaves the treatment plant. Slaked lime (calcium hydroxide) is used to neutralise the acid.

$$Cl_2 \ + \ H_2O \ \rightarrow \ HCl \ + \ HClO$$
$$2HCl \ + \ 2HClO \ + \ 2Ca(OH)_2 \ \rightarrow \ CaCl_2 \ + \ Ca(OCl)_2 \ + \ 4H_2O$$

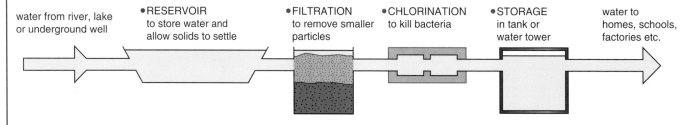

water from river, lake or underground well

• RESERVOIR
to store water and allow solids to settle

• FILTRATION
to remove smaller particles

• CHLORINATION
to kill bacteria

• STORAGE
in tank or water tower

water to homes, schools, factories etc.

Figure 6 Stages in the purification of water supplies

Pollution

A **pollutant** is a chemical, or too much of a chemical, in a situation where it is likely to cause harm. (Compare this to weeds that are plants growing where they are not required, a poppy is flower in a garden but a weed in a field of corn.) There are many ways in which water can be polluted. You need to concentrate on two: the use of fertilisers and of detergents.

● **Fertilisers** are used to get a better crop yield; they contain the nitrates and phosphates required for healthy growth of plants. Unfortunately, if too much is used the excess can be run off from the fields, by rain, into the water system.

● **Detergents** containing phosphates as brighteners also contribute to water pollution.

The nitrates and phosphates cause the water plants, including algae, to grow more vigorously, but, as they do so they use up the available oxygen. The reduction in the amount of oxygen in the water seriously affects the animal life in the water. With serious pollution all, or nearly all, the plants and animals will die. Without enough oxygen the dead organisms do not decay completely and smelly remains are left in the water. This process is called **eutrophication**.

Our drinking water can be contaminated by the presence of nitrates if they are washed into rivers, lakes and other bodies of water that are used as a source of water for human consumption.

Websites

www.soton.ac.uk/~engenvir/environment/water/water.html

www.gcsechemistry.com/ukop.htm

www.wildlifetrust.wg.uk/facts/rivers.htm

Exam questions

1 a) Describe a fair test you could carry out to compare the hardness of mineral water with the hardness of tap water. Give clear details of your method describing what you would expect to happen.

(3 marks)

   b) Hard water can be softened by ion exchange. Explain how ion exchange works.

(2 marks)

2 Here is a picture of an underground cave found in a limestone region.

a) Give the name of **one** of the icicle-like objects shown.

(1 mark)

b) What type of water is found in limestone regions?

(1 mark)

c) Why might kettles used to boil water in limestone regions need to be replaced more often than in other regions?

(1 mark)

3 Water is the most important solvent known to man.

a) Give **one chemical** test for water and the result you would expect to observe if water was present.

(3 marks)

b) Public water supplies go through two **main** processes to convert reservoir water to drinking water.

(i) Copy the diagram below and write the name of each process in the blank boxes.

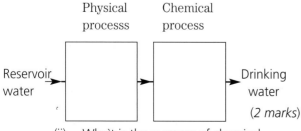

(ii) What is the purpose of chemical process?

(1 mark)

c) A student tested four samples of water (each 5 cm³) from different areas of Northern Ireland by shaking with 3 drops of soap solution. Another 5 cm³ of each sample of water was boiled and the soap solution test was repeated. The observations were recorded in the table.

Sample	Observation with soap solution	Observation for boiled sample with soap solution
A	No lather	Lather
B	Lather	Lather
C	Lather	Lather
D	No lather	No lather

(i) Which **two** samples contain hard water?

(2 marks)

(ii) Which **one** sample contains temporary hard water?

(1 mark)

(iii) Explain how you worked out your answer to part c)(ii).

(2 marks)

(iv) Explain how temporary hardness arises in rainwater.

(3 marks)

(v) Give a balanced symbol equation for the reaction in part c)(iv).

(2 marks)

d) The sample of permanent hard water may be softened using washing soda.

(i) Explain, in terms of ions, how permanent hard water is softened using washing soda.

(3 marks)

(ii) Give a balanced symbol equation for the reaction in part d)(i).

(2 marks)

e) Give two advantages and two disadvantages of hardness in a town's water supply.

(4 marks)

Chapter 10

Electrolysis

Learning objectives

By the end of this chapter you will:

➤ Know how to explain simply what happens in electrolysis

➤ Know about the importance of ions in electrolysis

➤ Understand the meanings of the terms anode, cathode and electrolyte

➤ Know of the use of electrolysis in the extraction of aluminium and in the refining of copper

➤ Know of the use of electrolysis in the chlor-alkali industry

➤ Know how to predict the products of electrolysis and how to write equations for the processes happening at the electrodes

Some definitions

Electrolysis is the passage of an electric current through a liquid that causes a chemical reaction.

The liquid, which is called the **electrolyte**, is an ionic compound that is **molten** (has been melted) or dissolved in water.

The reaction is a **decomposition** (breakdown) and, because it is caused by an electric current, electrolysis is also known as **electrolytic decomposition**.

Metal strips or graphite rods are placed in the liquid and are connected to a supply of electricity. The strips or rods are known as **electrodes**. The electrode attached to the negative pole is the **cathode** and the electrode attached to the positive pole is the **anode**. The electrodes may or may not take part in the reaction; if they do not they are called **inert electrodes**.

The negative cathode attracts positive ions, which are called **cations**.

The positive anode attracts negative ions, which are called **anions**.

As the ions move in different directions through the electrolyte it is decomposed and the electric current flows through the liquid.

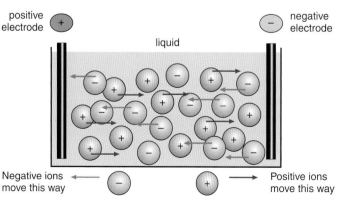

Figure 1 How the ions move in a liquid ionic compound

Products of electrolysis

Molten electrolytes

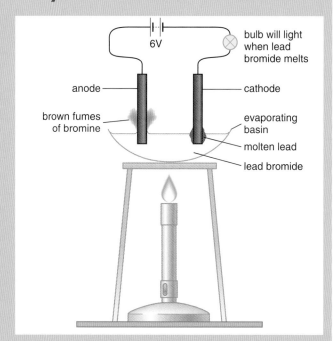

Figure 2 Electrolysis of a molten electrolyte

Table 1 The electrolysis of molten salts I

Electrolyte	Cathode product	Anode product
Molten lead bromide	lead	bromine
Molten lithium chloride	lithium	chlorine

The reactions at the anodes can be represented by ionic half equations.

Molten lead bromide:
Cathode: $Pb^{2+}(l) + 2e^- \rightarrow Pb(s)$
Anode: $2Br^-(l) \rightarrow Br_2(g) + 2e^-$
Overall reaction: $PbBr_2(l) \rightarrow Pb(l) + Br_2(g)$

Molten lithium chloride:
Cathode: $Li^+(l) + e^- \rightarrow Li(l)$
Anode: $2Cl^-(l) \rightarrow Cl_2(g) + 2e^-$
Overall reaction: $2LiCl(l) \rightarrow 2Li(l) + Cl_2(g)$

Predicting the products of electrolysis of a molten electrolyte

The only compound present is the electrolyte and so, with inert electrodes there are only two possible products. The cation is discharged at the cathode and the anion at the anode.

Table 2 The electrolysis of molten salts II

Molten electrolyte	Cathode product	Anode product
Calcium chloride	calcium	chlorine
Magnesium bromide	magnesium	bromine
Zinc oxide	zinc	oxygen

Electrolytes in aqueous solution

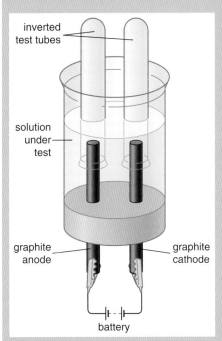

inverted test tubes

solution under test

graphite anode

graphite cathode

battery

Figure 3 Investigating the products when solutions are electrolysed

Water can ionise to give the hydrogen cation, H^+, and the hydroxide anion, OH^-.

$$H_2O(l) \rightarrow H^+(aq) + OH^-(aq)$$

These ions can be discharged at the appropriate electrodes

Cathode: $2H^+(aq) + 2e^- \rightarrow H_2(g)$
Anode: $4OH^-(aq) \rightarrow 2H_2O(l) + O_2(g)$

This means that there are two cations and two anions in the solution. If the electrolyte is a metal compound in solution then the cathode product is hydrogen unless the metal, for example copper, is less reactive than hydrogen.

If the electrolyte solution is dilute the anode product is most likely to be oxygen from the hydroxide but a concentrated solution may result in the anion forming the anode product.

Table 3 The electrolysis of aqueous solutions I

Electrolyte	Cathode product	Anode product
Dil. sulphuric acid	hydrogen	oxygen
Conc. sodium chloride	hydrogen	chlorine

With dilute sulphuric acid the only cation present is hydrogen (from both the acid and the water). Of the two anions, sulphate and hydroxide, the sulphate is much more reactive and so it stays in solution.

With concentrated sodium chloride solution the two cations are sodium and hydrogen. Sodium is much more reactive than hydrogen and so the sodium stays in solution. As the solution is concentrated chlorine is produced at the anode.

Predicting the products of electrolysis of an aqueous electrolyte

There is competition between the ions to remain in solution; in other words, to remain in the more stable state. The rule is that the more reactive substance stays in solution and the less reactive is discharged.

Cations: in order of staying in solution
$$K^+ > Na^+ > Ca^{2+} > Mg^{2+} > Al^{3+} > Zn^{2+} > Fe^{2+/3+} > H^+ > Cu^{2+}$$

Anions: in order of staying in the solution
$$SO_4^{2-} > Cl^- > Br^- > OH^- > I^-$$

In some cases the concentration of the solution has an effect on the nature of the anode product as in the concentrated sodium chloride solution referred to earlier.

Table 4 The electrolysis of aqueous solutions II

Aqueous electrolyte	Cathode product	Anode product
Dil. magnesium chloride	hydrogen	oxygen
Dil. copper sulphate	copper	oxygen
Conc. zinc bromide	hydrogen	bromine

Questions

1 Give the cathode and anode product for each of the following electrolytes:

a) molten potassium iodide
b) concentrated copper chloride solution
c) dilute sodium sulphate solution
d) molten calcium oxide
e) molten magnesium bromide

2 Write balanced ionic equations for the cathode and anode reaction for each of the following electrolytes:

a) dilute zinc sulphate solution
b) concentrated lithium chloride solution
c) molten lead fluoride
d) dilute potassium iodide
e) molten silver oxide

Electrolysis and redox

Electrolysis is a redox reaction.
Electrons are always lost at the anode and so all anode reactions are oxidations.

Example: $2Br^- \rightarrow Br_2 + 2e^-$

The bromide is oxidised to bromine.

Electrons are always gained at the cathode and so cathode reactions are always reductions.

Example: $Cu^{2+} + 2e^- \rightarrow Cu$

The copper ion is reduced to the copper atom.

Electrolysis in industry

Extraction of aluminium

Aluminium is a very reactive metal and is not easily extracted from its ore by chemical means.

Electrolysis can be used but only where there is a cheap supply of electricity. This is why aluminium extraction plants are sited where hydroelectric power is available. The main ore of aluminium is bauxite, it contains hydrated aluminium oxide. The ore is crushed, dried and purified. Then the aluminium oxide is dissolved in molten cryolite. The cryolite lowers the melting point of the aluminium oxide (thus decreasing the cost of the process) but does not affect the electrolysis of the aluminium oxide.

The electrolysis takes place in a steel tank lined with carbon. This lining acts as the cathode and carbon rods dipping into the molten mixture are the anodes.

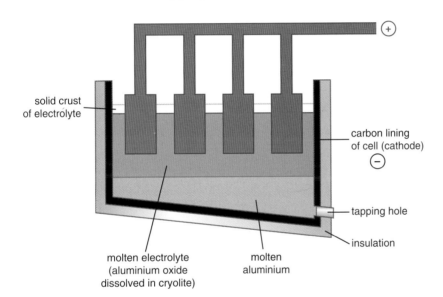

solid crust of electrolyte

carbon lining of cell (cathode)

tapping hole

insulation

molten electrolyte (aluminium oxide dissolved in cryolite)

molten aluminium

Figure 4 The electrolytic cell for aluminium manufacture

Molten aluminium is deposited at the cathode and sinks to the bottom of the tank from where it is tapped off from the cell.

$$Al^{3+} + 3e^- \rightarrow Al$$

The aluminium is reduced in this process, it gains electrons.

Oxygen is given off at the anodes.

$$2O^{2-} \rightarrow O_2 + 4e^-$$

At the temperature of the electrolysis the oxygen reacts with the carbon of the electrodes forming carbon dioxide.

$$C + O_2 \rightarrow CO_2$$

As a result the anodes constantly have to be replaced.

Refining of copper

Although copper is fairly easily extracted from its ore by chemical means, if very pure metal is required a further process is necessary. Electrolysis can be used to produce this very pure copper.

The copper ions in the electrolyte move to the cathode. They gain electrons forming copper atoms that are deposited on the pure copper cathode. As a result the cathode increases in size.

$$Cu^{2+}(aq) + 2e^- \rightarrow Cu(s)$$

At the anode the copper atoms lose electrons and become ions. The ions leave the anode and enter the electrolyte. The anode gets smaller.

$$Cu(s) \rightarrow Cu^{2+}(aq) + 2e^-$$

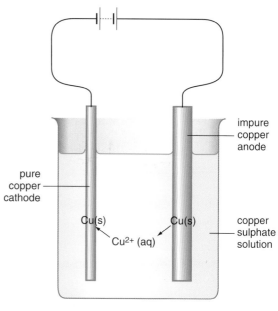

Figure 5 Refining of copper (laboratory version)

The impurities in the copper of the anode include silver and gold. As the anode decreases in size these impurities fall to bottom of the electrochemical cell as a sludge. This sludge is removed from the electrochemical cell in order to obtain the silver and gold.

Figure 6 Impure copper anodes being transferred to an electrolysis tank for purification

Electrolysis of brine (the chlor–alkali process)

Brine (sodium chloride solution) is electrolysed to produce chlorine, hydrogen and sodium hydroxide.

The ions present in the solution are Na^+, Cl^-, H^+ and OH^-

The chloride ions are discharged at the anode:

$$2Cl^- \rightarrow Cl_2 + 2e^-$$

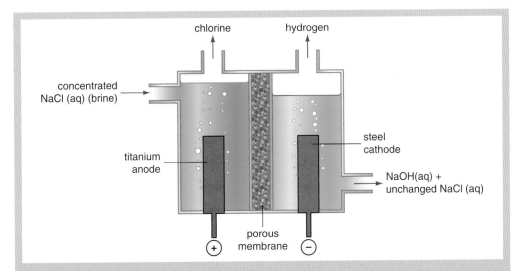

Figure 7
Manufacturing chlorine, sodium hydroxide and hydrogen using a membrane cell

The hydrogen ions are discharged at the cathode:

$$2H^+ + 2e^- \rightarrow H_2$$

The remaining ions are Na^+ and OH^-, they form the sodium hydroxide.

Although the electricity used is expensive the salt is cheap and all three of the products have further industrial uses.

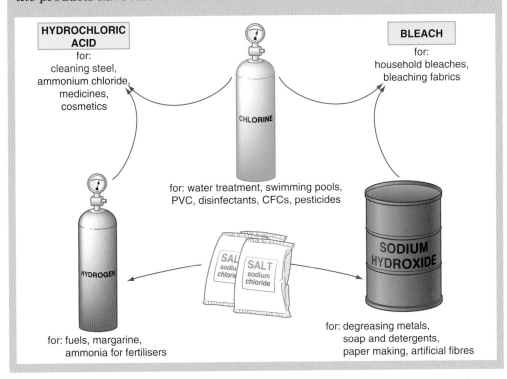

Figure 8 Important products from the chlor-alkali process

Websites

www.gcsechemistry.com/ukop.htm

www.s-cool.co.uk

www.schoolscience.co.uk

Exam questions

1 The diagram below shows the electrolysis of lead(II) bromide.

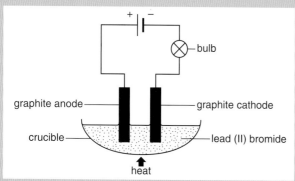

a) Why must lead(II) bromide be molten before the bulb will light up?
(*1 mark*)

b) How does the flow of current in an electrolyte differ from the flow of current in a metal? (*2 marks*)

c) What will be observed at the anode during the electrolysis of lead(II) bromide? (*1 mark*)

d) Write a balanced ionic equation to show the reaction taking place at the cathode. (*1 mark*)

2 Chlorine, hydrogen and sodium hydroxide are produced during the electrolysis of concentrated sodium chloride solution, which can be carried out as shown in the diagram below.

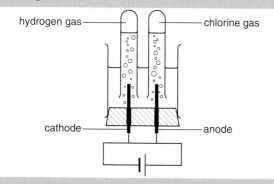

a) Why should care be taken to ensure that there are no naked flames near the apparatus? (*1 mark*)

b) How could you show that the solution remaining at the end of the electrolysis contains sodium hydroxide? (*1 mark*)

c) Explain why the solution remaining at the end contains sodium hydroxide.
(*2 marks*)

3 a) A student carried out an experiment using the apparatus shown in the diagram.

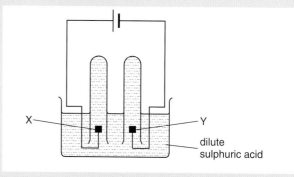

(i) What is the name given to
[A] electrode X? (*1 mark*)
[B] electrode Y? (*1 mark*)
[C] the conducting solution of acid?
(*1 mark*)

(ii) Name the gas formed at
[A] electrode X (*1 mark*)
[B] electrode Y (*1 mark*)

(iii) At which electrode will the greater volume of gas be produced?
(*1 mark*)

(iv) Name a substance from which the electrodes could be made.
(*1 mark*)

(v) Give the balanced, ionic equations for the reactions taking place at the electrodes.
(*4 marks*)

b) The student repeated the experiment using concentrated sodium chloride solution in the beaker.

(i) Name the product formed at the
[A] positive electrode. (*1 mark*)
[B] negative electrode. (*1 mark*)

(ii) Give the balanced, ionic equation for the reaction taking place at the positive electrode.
(*2 marks*)

4 Electrolysis plays an important part in extracting some metals from their ores and in purifying others.

a) Define what is meant by the term electrolysis.

(*2 marks*)

b) Name **one** metal that is obtained from its ore by electrolysis.

(*1 mark*)

c) Draw a **labelled** diagram of simple laboratory apparatus which shows how copper can be **purified**.

(*6 marks*)

d) Write **balanced, ionic** equations to show the electrode reactions which occur when copper is purified.

(i) At the anode (*2 marks*)

(ii) At the cathode (*2 marks*)

e) Which particles are responsible for the conductivity of electricity in

(i) metals? (*1 mark*)

(ii) electrolytes? (*1 mark*)

f) Give **two** uses of copper metal apart from its use in electrical wiring.

(*2 marks*)

g) The electrolysis of dilute sulphuric acid using inert electrodes produces two gases as products.

(i) Name a material which could be used as the electrodes in this experiment.

(*1 mark*)

(ii) Name the product formed at the

[A] anode. (*1 mark*)

[B] cathode. (*1 mark*)

(iii) Write **balanced, ionic** equations to show how these gases are formed.

[A] Anode reaction. (*2 marks*)

[B] Cathode reaction. (*2 marks*)

5 Aluminium metal is extracted from pure aluminium oxide by electrolysis using the cell shown below.

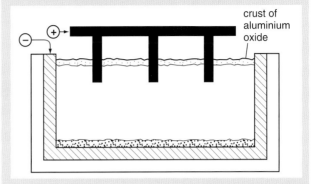

a) Name the substance in which aluminium oxide is dissolved in this process.

(*1 mark*)

b) Explain why the substance in part a) is added. (*3 marks*)

c) Name the substance from which both electrodes are made. (*1 mark*)

d) At what temperature does the process take place? (*1 mark*)

e) Give a balanced, ionic equation for the reaction which takes place at the **negative** electrode.

(*2 marks*)

f) Is the reaction in part e) an oxidation or reduction? Explain your answer.

(*3 marks*)

g) How is the product formed at the negative electrode removed from the cell? (*2 marks*)

h) Describe what happens at the positive electrode during electrolysis.

(*2 marks*)

i) Give a balanced, ionic equation for the reaction in part h).

(*2 marks*)

j) Which electrode has to be frequently replaced during the process? Explain your answer. (*2 marks*)

Chapter 11

The Periodic Table

By the end of this chapter you should be able to:

➤ Outline the work of *Newlands* and *Mendeleev* in the development of the Periodic Table

➤ Relate the position of the first twenty elements in the Periodic Table to their electronic structure

➤ Demonstrate a knowledge that the Periodic Table groups together elements with similar properties, for example, alkali metals as a group of reactive metals, halogens as a group of reactive non-metals, the noble gases as a group of unreactive non-metals

➤ Describe simple trends in the properties of elements within groups (I, II and VII) and across periods (2 and 3)

➤ Use the Periodic Table to predict the properties of certain unfamiliar elements, limited to groups I, II and VII and properties to relative atomic mass, atomic size, metallic and non-metallic characteristics, valency and chemical reactivity with oxygen, water and dilute acids

➤ Explain the uses and limitations of different systems of classification, for example, metals and non-metals; limited to silicon and germanium as semi-metals and to electrical conduction in graphite

Newlands and Mendeleev

Around the mid-nineteenth century many chemists were researching and publishing information on the properties of elements. More and more they realised that it was necessary to bring some type of order to the huge amount of information that was available on the already discovered elements. There was considerable attention given to the idea that elements could be arranged in terms of their properties and atomic weights (now called relative atomic masses.)

In 1866, an English chemist, John Newlands, published the **Law of Octaves**, where he ordered the known elements in terms of their increasing relative atomic mass. Newlands proposed that the eighth element was a repetition of the first, just like the eight notes of an octave of music. Thus starting from lithium, Li, and moving on eight places we arrive at sodium, Na, an element with very similar properties to lithium. In the same way we could show this eighth relationship for a number of other elements in Newlands' arrangement.

H	Li	Be	B	C	N	O
F	Na	Mg	Al	Si	P	S
Cl	K	Ca	Cr	Ti	Mn	Fe
Co and Ni	Cu	Zn	Y	In	As	Se
Br	Rb	Sr				

Newlands' arrangement of some of the elements in octaves is shown above.

Although Newlands had made a significant contribution to the classification of the elements, his work was rejected by other scientists because:

● His classification assumed that all elements had been discovered, even though a number of new elements had been recognised a few years prior to his publication. No gaps were left for undiscovered elements. The discovery of new elements would immediately destroy his arrangement of octaves.

● Newlands grouped some elements together which were not alike. For example, manganese was grouped with nitrogen, phosphorus and arsenic while copper was grouped with lithium, sodium, potassium and rubidium.

● In forming the octaves, Newlands found it necessary to place two elements into one position as can be seen for cobalt and nickel in the table on page 106.

Although Newlands had identified important periodic properties between certain elements based on his Law of Octaves, it was the Russian chemist Dmitri Mendeleev in 1869 who received the main credit for arranging the elements in the Periodic Table. Like Newlands he arranged the elements in order of their increasing relative atomic mass. He also placed elements with similar properties in the same vertical column as shown in Figure 1a, however, Mendeleev realised that there were elements that had not been discovered and he left gaps for them. The gaps also helped him keep similar elements in the same vertical columns. The vertical columns of elements were called **Groups** and the horizontal rows of elements were called **Periods**.

	GROUP							
	I	II	III	IV	V	VI	VII	VIII
Period 1	H							
Period 2	Li	Be	B	C	N	O	F	
Period 3	Na	Mg	Al	Si	P	S	Cl	
Period 4	K Cu	Ca Zn	* *	Ti *	V As	Cr Se	Mn Br	Fe Co Ni
Period 5	Rb Ag	Sr Cd	Y In	Zr Sn	Nb Sb	Mo Te	* I	Ru Rh Pd

Figure 1 Part of Mendeleev's Periodic Table

Mendeleev predicted that, through time, elements would be discovered that would fill the gaps he had left in his Periodic Table. He also predicted the properties of many unknown elements with amazing accuracy; for example, in the gaps below aluminium and silicon he predicted the properties of two unknown elements, eka-aluminium and eka-silicon. Later when the two elements were discovered (1875 and 1886 respectively) they were called gallium and germanium and had properties almost identical to those predicted by Mendeleev. Confidence in the Periodic Table soared.

The accuracy of Mendeleev's predictions influenced other scientists and soon it became accepted that the Periodic Table was a suitable way of ordering the elements and their properties.

The Modern Periodic Table

The modern Periodic Table, Figure 3 is based on that of Mendeleev, however, there are a number of changes:

● Mendeleev placed the elements in order of relative atomic mass. However, in the modern Periodic Table elements are placed in order of their atomic number. When Mendeleev placed elements in order of relative atomic mass he found that in some instances he had to reverse the order of elements, for example, tellurium, Te = 128 and iodine, I = 127. According to ordering by relative atomic mass, Mendeleev should have placed iodine before tellurium. However, he realised that for iodine and tellurium to fit into their proper groups he had to reverse the order. This problem is now overcome using atomic number where tellurium = 52 and iodine = 53.

Figure 2
Dmitri Mendeleev

Group Period	1 Alkali metals	2 Alkaline -earth metals										3	4	5	6	7 Halogens	0 Noble gases
1				**Key**													He helium 2
2	Li lithium 3	Be beryllium 4										B boron 5	C carbon 6	N nitrogen 7	O oxygen 8	F flourine 9	Ne neon 10
3	Na sodium 11	Mg magnesium 12				transition elements						Al aluminium 13	Si silicon 14	P phosphorus 15	S sulphur 16	Cl chlorine 17	Ar argon 18
4	K potassium 19	Ca calcium 20	Sc 21	Ti 22	V 23	Cr chromium 24	Mn manganese 25	Fe iron 26	Co 27	Ni 28	Cu copper 29	Zn zinc 30	Ga 31	Ge 32	As 33	Se 34	Br bromine 35	Kr krypton 36
5	Rb 37	Sr 38	Y 39	Zr 40	Nb 41	Mo 42	Tc 43	Ru 44	Rh 45	Pd 46	Ag silver 47	Cd 48	In 49	Sn tin 50	Sb 51	Te 52	I iodine 53	Xe 54
6	Cs 55	Ba 56	57-71 see below	Hf 72	Ta 73	W 74	Re 75	Os 76	Ir 77	Pt platinum 78	Au gold 79	Hg mercury 80	Tl 81	Pb lead 82	Bi 83	Po 84	At 85	Rn 86
7	Fr 87	Ra 88	89-103 see below	Ku 104	Ha 105	106	107	108	Mt 109									

Key:
H
hydrogen
1
← symbol
← name
← atomic number

H
hydrogen
1

	lanthanides	La lanthanum 57	Ce 58	Pr 59	Nd 60	Pm 61	Sm 62	Eu 63	Gd 64	Tb 65	Dy 66	Ho 67	Er 68	Tm 69	Yb 70	Lu 71
	actinides	Ac actinium 89	Th 90	Pa 91	U uranium 92	Np 93	Pu 94	Am 95	Cm 96	Bk 97	Cf 98	Es 99	Fm 100	Md 101	No 102	Lr 103

Figure 3 Modern Periodic Table

- A family of very unreactive elements, called the noble gases or Group 0 elements, has been discovered and inserted.
- The transition metals have been taken out and placed as a block of metals between Group II and Group III.
- Some of the groups have been given common names as well as group numbers:

Table 1 Group names in the Periodic Table

Group number	Name	Elements
I	Alkali metals	Lithium, sodium, potassium, rubidium, caesium and francium
II	Alkaline earth metals	Beryllium, magnesium, calcium, strontium, barium and radium
VII	Halogens	Fluorine, chlorine, bromine, iodine and astatine
0	Noble gases	Helium, neon, argon, krypton, xenon and radon

- There is a clear distinction between metals and non-metals. The non-metals are found at the right hand side of Figure 3, with the thick line acting as a division between metals and non-metals.

Along the steps there are some elements that are classified as semi-metals. Semi-metals are elements with properties between those of metals and non-metals. They generally look like metals but are brittle as non-metals are. One example of such intermediate behaviour is the fact that they are semi-conductors, and their electrical conduction properties are between those of metals and non-metals. Silicon and germanium are two examples of semi-conductors. Graphite, which is a form of carbon, is an unusual non-metal in that it is a good conductor of electricity, a property typical of metals. This property is explained by the structure of graphite which has a mobile cloud of free electrons in the layers.

Trends in the Periodic Table

We have already seen that as we move across a period the properties of the elements gradually change from metals through to semi-metals to non-metals. This trend and some other properties of the elements in Period 3 are given below:

Property	Sodium Na	Magnesium Mg	Aluminium Al	Silicon Si	Phosphorus P	Sulphur S	Chlorine Cl	Argon Ar
Electronic structure	2.8.1	2.8.2	2.8.3	2.8.4	2.8.5	2.8.6	2.8.7	2.8.8
Group	I	II	III	IV	V	VI	VII	VIII
Valency	1	2	3	4	3 or 5	2	1	0
Metal/ Non-metal	Metal	Metal	Metal	Semi-metal	Non-metal	Non-metal	Non-metal	Non-metal
Structure	Giant metallic	Giant metallic	Giant metallic	Giant covalent	Simple molecular	Simple molecular	Simple molecular	Atomic
Bonding	Metallic	Metallic	Metallic	Covalent	Covalent	Covalent	Covalent	Single atoms
State	Solid	Solid	Solid	Solid	Solid	Solid	Gas	Gas
Melting point °C	98	650	659	1410	44	119	−101	−189
Oxide	Basic	Basic	Amphoteric	Acidic	Acidic	Acidic	Acid	−

Table 2

From the table it is seen that

- The period number gives the number of shells of electrons which atoms have in that period. For example, as sodium is in Period 3 it has three shells of electrons, with two electrons in the first shell, eight in the second and one in the third and its electronic structure is 2,8,1.

- The number of electrons in the outer shell of an element is always the same as the group number, for example, sodium has one electron in its outer shell and belongs to Group I while chlorine has seven electrons in its outer shell and belongs to Group VII. Thus elements in the same group will always have the same number of electrons in their outer shell. As it is the number of electrons in the outer shell of an atom that dictates its chemical properties, then elements in the same group will have similar chemical properties. For example, Group 1 metals, known as the alkali metals, are a group of very reactive metals, Group VII elements, the halogens, are a group of very reactive non-metals while Group 0 elements, the noble gases, with filled outer shells of electrons, are a group of very unreactive non-metals.

- Moving across a period results in the outer shell being filled with electrons. In Period 3, the third shell is progressively filled in going from sodium 2,8,1 to argon 2,8,8.

- The size of atoms decreases going across a period but increases going down a group. In Period 3 we can predict that a sodium atom will be larger than a sulphur atom. If we go down group 1 metals, it is possible to predict that a caesium atom will be bigger than a sodium atom.

- For Groups I and II metals the reactivity increases going down the group, the most reactive metal being Francium at the bottom of Group I.

- For the halogens, the reactivity increases going up the group. Fluorine is the most reactive non-metal.

- Melting points increase from sodium (98°C) to silicon (1410°C) and then drop sharply to argon (-189°C). This can be explained by the change in structure moving across the period with giant metallic structure (sodium) at the left hand side, giant covalent (silicon) in the middle of the period and simple molecular (chlorine) at the right hand side. This also corresponds to a change from metals through to semi-metals to non-metals.

Questions

1 Copy this paragraph and fill in the blanks.

Perhaps the greatest contributor to the development of the Periodic Table was the Russian scientist _____ in 1869. He stated 'when elements are arranged in order of increasing _____ similar properties recur at intervals'. He wrote down what he knew about each element on a separate card and then sorted the cards into 'piles of elements' with similar _____. His inspiration was to leave gaps for _____ elements. Nowadays, the elements are arranged in order of increasing _____ .

2 From your Periodic Table give the symbol for an element which is:

   a) a noble gas

   b) an alkaline earth metal

   c) a halogen

   d) an alkali metal.

3 Use your Periodic Table to name:

   a) a metal in Period 3 with 3 electrons in its outer shell

   b) a non-metal in Period 2 which belongs to Group VII

   c) a noble gas with 2 electrons in its outer shell

   d) an element which forms a stable ion by gaining 2 electrons.

4 Give three features of the Periodic Table developed by Mendeleev.

5 Use the following list of elements to answer the questions below:

   phosphorus, sulphur, aluminium, sodium, magnesium, chlorine and potassium

   a) Select the element which has the largest atomic size and the element which has the smallest atomic size

   b) give the symbol for the most reactive metal in the list

   c) name the element which forms diatomic molecules

   d) what type of bonding exists in sodium?

6 Write a poem or limerick which illustrates some of the properties of an element from either Group I or Group II. The following example about chlorine should help you.

c
h
l
o
r
i
n
e

Even though this **halogen** is **green**
Chlorine can hardly be seen
It can do you much harm
So always raise the alarm
Cause this **toxic gas** is so mean

c
h
l
o
r
i
n
e

Word process your verse and add drawings or pictures to personalise your work.

Trends in the properties of Groups I, II and VII elements

Group I: alkali metals

The Group I or alkali metals is a group of very reactive metals. Due to their high reactivity they must be stored under oil to prevent them from reacting with oxygen or water vapour in the air. The metals are all soft metals and when cut they have a silvery lustre but soon tarnish in the air. Although they all are very reactive, the reactivity of the metals increases down the group. Owing to their high reactivity, great care must be taken when using the alkali metals. You may study the reactions of lithium, sodium and potassium with water but you will not use rubidium or caesium because they are too reactive and explode violently in water. Francium, the heaviest alkali metal does not occur naturally and is radioactive. The most stable isotope of francium is $^{223}_{87}$Fr which has a half-life of 20 minutes.

Physical properties of the alkali metals

From the physical properties of lithium, sodium and potassium in Table 3 it is seen that:

● The metals have low melting points.

● The melting and boiling points decrease going down the group.

● They are soft metals that can be cut with a knife.

● The metals have a low density and all three float on water.

Element	Atomic number	Relative atomic mass	Melting point (°C)	Boiling point (°C)	Density (g/cm³)
lithium	3	7	180	1330	0.53
sodium	11	23	98	883	0.97
potassium	19	39	64	760	0.86

Table 3 Physical properties of lithium, sodium and potassium

Chemical reactions of the alkali metals

Reaction with water

All the alkali metals react vigorously with water to produce hydrogen and an alkaline solution of the metal hydroxide (see Figure 4). For Li, Na and K all three metals float on the surface of the water, melt and as they rapidly move around often break into flame. Moving down the group the reactions become more vigorous.

a)

b)

c)

Figure 4 The reactions of a) lithium, b) sodium and c) potassium with water

$$\text{lithium} + \text{water} \rightarrow \text{lithium hydroxide} + \text{hydrogen}$$
$$2Li(s) + 2H_2O(l) \rightarrow 2LiOH(aq) + H_2(g)$$

Lithium reacts with water forming a steady stream of hydrogen gas and an alkaline solution of lithium hydroxide.

$$\text{sodium} + \text{water} \rightarrow \text{sodium hydroxide} + \text{hydrogen}$$
$$2Na(s) + 2H_2O(l) \rightarrow 2NaOH(aq) + H_2(g)$$

Sodium reacts vigorously with water producing hydrogen gas. Sometimes the hydrogen gas ignites and burns with a bright yellow flame because the reaction is highly exothermic.

$$\text{potassium} + \text{water} \rightarrow \text{potassium hydroxide} + \text{hydrogen}$$
$$2K(s) + 2H_2O(l) \rightarrow 2KOH(aq) + H_2(g)$$

Potassium reacts violently and the hydrogen ignites immediately because the reaction is very fast and highly exothermic.

Reaction with air/oxygen

When lithium, sodium and potassium are exposed to air they tarnish and form a layer of the metal oxide. The rate at which they tarnish increases as the group is descended. Lithium tarnishes slowly, potassium tarnishes very rapidly and sodium tarnishes fairly quickly. The equation below shows the reaction taking place when lithium burns in oxygen with a deep red flame to form lithium oxide, Li_2O.

$$4Li(s) + O_2(g) \rightarrow 2Li_2O(s)$$

When sodium and potassium are heated in air they burn to form metal oxides. Sodium burns with a yellow flame to produce a mixture of sodium oxide, Na_2O, and sodium peroxide, Na_2O_2; while potassium burns with a lilac flame to form potassium superoxide, KO_2.

$$2Na(s) + O_2(g) \rightarrow Na_2O_2(s)$$

$$K(s) + O_2(g) \rightarrow KO_2(s)$$

Group II: alkaline earth metals

Like Group I metals, the alkaline earth metals are a group of reactive metals that lie to the left hand side of the Periodic Table. Group II metals are harder and more dense than the corresponding metals in Group I. From the table below it is seen that the melting and boiling points and densities of Group II metals are higher than those in Group I. The reason for the differences in these properties is due to the stronger metallic bonding in Group II metals. Each atom has two valence or outer electrons rather than one to donate to the mobile cloud of electrons and this results in stronger metallic bonding (see page 23).

Table 4 Properties of Group II metals

Metal	Atomic number	Relative atomic mass	Melting point (°C)	Boiling point (°C)	Density (g/cm³)
Beryllium (Be)	4	9	1283	2487	1.8
Magnesium (Mg)	12	24	650	1117	1.7
Calcium (Ca)	20	40	850	1492	1.6
Strontium (Sr)	38	88	769	1384	2.5
Barium (Ba)	56	137	725	1640	3.6
Radium (Ra)	88	226	700	1140	5.0

As for Group I metals, the chemical reactivity of the Group II metals increases down the group; beryllium is the least reactive metal and barium is the most reactive. However, Group II metals are not as reactive as those of Group I, but, barium, like group 1 metals, reacts readily with oxygen and water vapour in the atmosphere and must be stored under oil. Like francium, radium, the last element in Group II is radioactive.

Chemical reactions of the alkaline earth metals

Reaction with air/oxygen

The alkaline earth metals burn in oxygen or air forming metal oxides. Magnesium burns with a brilliant white flame forming a white ash of magnesium oxide. Due to the intensity of this flame magnesium is used for flares and in fireworks. Calcium burns with a brick red flame forming the white solid calcium oxide while barium burns with an apple green flame giving barium oxide:

$$\text{magnesium} + \text{oxygen} \rightarrow \text{magnesium oxide}$$
$$2Mg(s) + O_2(g) \rightarrow 2MgO(s)$$
$$\text{calcium} + \text{oxygen} \rightarrow \text{calcium oxide}$$
$$2Ca(s) + O_2(g) \rightarrow 2CaO(s)$$

Reaction with water

The reaction with water increases markedly going down the group. Magnesium only reacts very slowly with cold water and it is difficult to observe the hydrogen given off. However, magnesium reacts much more quickly with steam producing magnesium oxide and hydrogen, (Figure 5):

$$\text{magnesium} + \text{steam} \rightarrow \text{magnesium oxide} + \text{hydrogen}$$
$$\text{Mg(s)} + \text{H}_2\text{O(g)} \rightarrow \text{MgO(s)} + \text{H}_2\text{(g)}$$

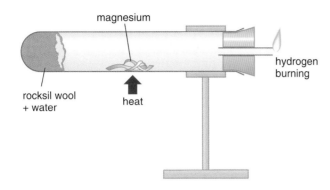

Figure 5 Magnesium reacting with steam

Calcium, like the Group I metals, reacts readily with cold water producing a steady stream of hydrogen gas and the sparingly soluble alkali, calcium hydroxide. The saturated solution of calcium hydroxide is known as limewater. Unlike the Group I metals, calcium does not float but sinks to the bottom of the water.

$$\text{calcium} + \text{water} \rightarrow \text{calcium hydroxide} + \text{hydrogen}$$
$$\text{Ca(s)} + 2\text{H}_2\text{O(l)} \rightarrow \text{Ca(OH)}_2\text{(s)} + \text{H}_2\text{(g)}$$

Strontium reacts more vigorously with cold water than calcium does, and barium even more so.

Reaction with acids

The alkaline earth metals react vigorously with dilute acids to produce salts and hydrogen, for example:

$$\text{Magnesium} + \text{hydrochloric acid} \rightarrow \text{magnesium chloride} + \text{hydrogen}$$
$$\text{Mg} + 2\text{HCl} \rightarrow \text{MgCl}_2 + \text{H}_2$$

$$\text{Magnesium} + \text{sulphuric acid} \rightarrow \text{magnesium sulphate} + \text{hydrogen}$$
$$\text{Mg} + \text{H}_2\text{SO4} \rightarrow \text{MgSO}_4 + \text{H}_2$$

Questions

7 This question is about the Group I metals

   a) Show how the electrons are arranged in an atom of sodium.
   b) Name the metal in Group I which has the smallest atom .
   c) Give the name of the most reactive metal in Group I.
   d) Why is potassium stored under oil?
   e) Why should caesium never be added to water?

8 A small piece of potassium was added to a trough of water
   a) Give four things you would observe in this reaction.
   b) write a symbol equation to show what happens when potassium is added to water.
   c) How could you show that the solution contained potassium hydroxide?

9 a) Write a symbol equation to show how lithium burns in air
   b) Complete the sentence 'lithium burns in air with a deep red flame while sodium burns with a _____ flame and potassium burns with a _____ flame'.
   c) Underline the metal that has the lowest melting point: lithium, sodium, potassium.

10 This question is about the elements in Group II:
   a) Show how the electrons are arranged in a calcium atom.
   b) How can you tell from the electronic structure that magnesium is a Group II metal?
   c) Give the formula of the stable ion formed by magnesium.
   d) Explain why Group II metals are harder and more dense than Group I metals.
   e) Name the Group II metal which reacts most vigorously when added to water.

11 This question is about calcium and magnesium:
   a) State three things you would observe if a small granule of calcium was added to a trough of cold water. Write a symbol equation for the reaction.
   b) Draw a diagram of the apparatus you would use to show that magnesium reacts with steam. Give two important safety precautions that must be taken when carrying out this reaction.
   c) Calcium burns in air with a reddish flame; write a symbol equation for this reaction.
   d) From your knowledge of Group II elements explain why magnesium is used in flares.
   e) Using electronic diagrams for magnesium and oxygen explain how the compound magnesium oxide forms. Name the type of bonding present and explain why magnesium oxide has a high melting point.

12 a) Write a symbol equation to show how magnesium reacts with dilute hydrochloric acid.
   b) Give formulae for the following Group II compounds: calcium carbonate, barium chloride, magnesium nitrate, magnesium hydrogencarbonate, calcium hydroxide and barium sulphate.
   c) Use your data leaflet to work out which of the compounds in 12 b) are soluble in water.

13 Use the website http://www.chemsoc.org to find out the following information about Group II elements:
   a) the date each element was discovered.
   b) the appearance of each element.
   c) the percentage abundance of each element in the earth's crust.
   Print off your information and present it to the class.

Group VII: the halogens

We are familiar with halogen chemistry in everyday life. Some common examples are:

- fluoride in toothpaste to reduce tooth decay
- salt (sodium chloride) in cooking and for de-icing roads
- iodine solution as a mild antiseptic
- the use of chlorine in water treatment
- the use of chlorine in PVC.

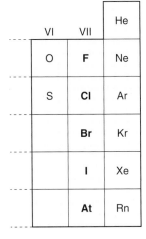

VI	VII	0
		He
O	F	Ne
S	Cl	Ar
	Br	Kr
	I	Xe
	At	Rn

Figure 6b Chlorine, bromine and iodine

Fluorine, chlorine, bromine and iodine are all elements that belong to a group of very reactive non-metals called the halogens. Due to their high reactivity they never occur free in nature but as part of a metal salt, for example, sodium chloride and magnesium bromide are both found in seawater. Due to the large number of salts which the halogens can form with metals, it is not surprising that the word halogen means 'salt former'.

The halogens or Group VII elements are positioned at the right side of the Periodic Table and they have seven electrons in their outer shell. The table below highlights the main physical properties of fluorine, chlorine, bromine and iodine:

Table 5 Physical properties of fluorine, chlorine, bromine and iodine

Element	Relative atomic mass	State at room temperature	Colour	Structure	Melting point (°C)	Boiling point (°C)
Fluorine	19.0	gas	pale yellow	F_2 molecules	−220	−188
Chlorine	35.5	gas	pale green	Cl_2 molecules	−101	−35
Bromine	79.9	liquid	red brown	Br_2 molecules	−7	58
Iodine	126.9	solid	dark grey	I_2 molecules	114	183

Physical properties and molecular structure

● Descending the group, melting and boiling points increase. Although the diatomic molecules have strong covalent bonds between the atoms, there are only weak attractive forces between the molecules. This explains the low boiling points and melting points of the halogens.

● Elements change state as the atomic number and relative atomic mass increases:

> chlorine (gas)   bromine (liquid)   iodine (solid)

● Moving down the group, the size of the molecules increases and this results in an increase in the attractive forces between the molecules. For iodine, the attractive forces are sufficiently strong to hold the molecules closely together in an orderly arrangement and as a result iodine is a solid. It is an unusual solid in that it sublimes on heating.

● Descending the group the elements change colour from yellow (fluorine) through green (chlorine) and red/brown (bromine) to dark grey/black (iodine).

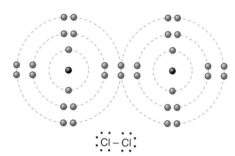

● All the elements have a simple covalent molecular structure where electrons are shared to obtain a stable electron structure to match that of the nearest noble gas. The diatomic structure of chlorine is shown in Figure 7.

● Because the halogens are covalent molecules they are poor conductors of heat and electricity.

Figure 7 A chlorine molecule, Cl_2

● As the atomic number increases the chemical reactivity decreases down the group (unlike Groups I and II).

Chemical reactions of the halogens

Displacement reactions

Unlike the metals in Groups I and II the reactivity of the non-metals decreases going down the group. Fluorine is the most reactive non-metal; chlorine is also very reactive, followed by bromine and then the moderately reactive iodine. The relative reactivity of the halogens can be shown using displacement reactions. Here a more reactive halogen is used to displace a less reactive halogen from a solution of its salt. Thus, when pale green chlorine gas is bubbled into colourless potassium bromide solution, the colour changes to a reddish hue due to the formation of bromine. In industry, this reaction is used in the production of bromine from sea water.

> chlorine + potassium bromide → bromine + potassium chloride
> (green)   (colourless)   (red)   (colourless)
> $Cl_2(g)$ + $2KBr(aq)$ → $Br_2(aq)$ + $2KCl(aq)$

Bubbling chlorine into potassium iodide causes the solution to turn a brownish colour due to the formation of iodine:

> chlorine + potassium iodide → iodine + potassium chloride
> (green)   (colourless)   (brown)   (colourless)
> $Cl_2(g)$ + $2KI(aq)$ → $I_2(aq)$ + $2KCl(aq)$

Reaction with water

Chlorine is moderately soluble in water and reacts to form a mixture of hydrochloric and chloric(I) acids. The resulting solution is known as 'chlorine water'.

$$\text{chlorine} + \text{water} \rightarrow \text{hydrochloric acid} + \text{chloric(I) acid}$$
$$Cl_2(g) + H_2O(l) \rightarrow HCl(aq) + HClO(aq)$$

Bromine also dissolves and reacts with water in a similar way to chlorine. If blue litmus paper is added to chlorine water it turns red due to the acids present and it then turns white as it gets bleached by chloric(I) acid.

Questions

14 This question is about the halogens:

a) To which group in the Periodic Table do the halogens belong?
b) Give one use of chlorine.
c) Describe two physical properties of bromine.
d) What would you observe if chlorine was bubbled into an aqueous solution of potassium iodide? Write a symbol equation for the reaction taking place.
e) Give a reason why the reaction in 14 d) must be carried out in a fume cupboard.

15 a) Draw a diagram to show how the electrons are arranged in a chlorine atom.
b) Use your diagram to explain how a chlorine molecule is formed.
c) Explain why chlorine gas has a low boiling point.
d) From your knowledge of the halogens explain why bromine is a liquid and iodine is a solid.
e) Explain why blue litmus paper, when added to chlorine water, first turns red and then white.

16 a) Draw diagrams to show how the electrons are arranged in atoms of lithium and fluorine. Use your diagrams to show how the compound lithium fluoride forms.
b) Give three properties of the compound lithium fluoride.

Websites

www.chemsoc.org/viselements/

www.webelements.com/index.html

Exam questions

1   This question is about some elements in the Periodic Table. You may find your Data Leaflet useful.

a) (i)   How many electrons has a silicon atom in its outer shell?

(1 mark)

(ii)   Give the common name for

[A] Group 7 elements.   *(1 mark)*

[B] Group 1 elements.   *(1 mark)*

(iii)   From **Period 3** name an element which is found in

[A] Group II?   *(1 mark)*

[B] Group V?   *(1 mark)*

b)   Argon is a noble gas used inside light bulbs.

(i)   Why is argon used inside light bulbs instead of air?

(1 mark)

(ii)   Show how the electrons of argon are arranged.

(1 mark)

(iii)   Why does the arrangement of electrons in argon make it suitable for use inside light bulbs?

(1 mark)

2   A portion of the Periodic Table of the Elements is shown below.

						H											He
Li	Be											B	C	N	O	F	Ne
Na	Mg											Al	Si	P	S	Cl	Ar
K	Ca	Sc	Ti	V	Cr	Mn	Fe	Co	Ni	Cu	Zn	Ga	Ge	As	Se	Br	Kr
Rb																	
Cs																	

Using the above portion of the Periodic Table answer questions a) to e).

a)   Give the **names** of

(i)   an alkali metal

(1 mark)

(ii)   a halogen

(1 mark)

(iii)   a noble gas

(1 mark)

b)   Give the **name** of the non-metal which is found in Group VI and Period 3.

(1 mark)

c)   In which **Group** of the Periodic Table would you find the

(i)   most reactive metals?

(1 mark)

(ii)   most reactive non-metals?

(1 mark)

d)   Give the **name** of a non-metal with a valency of 3.

(1 mark)

e)   **Name** the elements present in ammonia.

(2 marks)

3   Elements in the Periodic Table are arranged in groups and periods and in order of increasing atomic number.

a) (i)   Give the name and symbol of the element whose atomic number is **82**.

(2 marks)

(ii)   Give the period and the group in which you would find nitrogen.

(2 marks)

b)   Magnesium is a reactive metal.

(i)   Describe **two** observations you would make when magnesium burns in air.

(2 marks)

(ii)   Give **one** safety precaution which must be taken when magnesium is burned in air.

(1 mark)

(iii)   Magnesium will also burn in chlorine to give magnesium chloride. Write a symbol equation for this reaction.

(2 marks)

(iv)   Name a metal which is in the same group as magnesium but which is more reactive.

(1 mark)

c) Chlorine gas is a reactive non-metal.

   (i) What colour is chlorine gas?

      (1 mark)

   (ii) Give the **common** name of the group in which chlorine is found.

      (1 mark)

   (iii) Give **one** important source of chlorine.

      (1 mark)

   (iv) Name an element which is in the same group as chlorine but which is less reactive. Give the formula of its molecule.

      (2 marks)

4 a) Sodium is a Group I metal which reacts exothermically with water.

   (i) Give **two** safety precautions which **must** be taken when carrying out this reaction in a laboratory.

      (2 marks)

   (ii) What is the meaning of the term **exothermic**?

      (1 mark)

   (iii) What pH value would you expect for the solution formed when sodium reacts with water?

      (1 mark)

   (iv) Write a **word** equation for the reaction of sodium with water.

      (2 marks)

   (v) Name a Group I element which is **more** reactive than sodium.

      (1 mark)

5 The table below represents a number of the elements in the Periodic Table.

I	II			III	IV	V	VI	VII	O
		1 **H** 1 hydrogen							4 **He** 2 helium
7 **Li** 3 lithium	9 **Be** 4 beryllium			11 **B** 5 boron	12 **C** 6 carbon	14 **N** 7 nitrogen	16 **O** 8 oxygen	19 **F** 9 flourine	20 **Ne** 10 neon
23 **Na** 11 sodium	24 **Mg** 12 magnesium			27 **Al** 13 aluminium	28 **Si** 14 silicon	31 **P** 15 phosphorus	32 **S** 16 sulphur	35.5 **Cl** 17 chlorine	40 **Ar** 18 argon
39 **K** 19 potassium	40 **Ca** 20 calcium								

a) What name is given to the horizontal rows in the Periodic Table?

      (1 mark)

b) Name an element, from the table above, which has only one electron in its outer shell.

      (1 mark)

c) From the table above name the alkaline earth metal which reacts most vigorously with water.

      (1 mark)

d) When chlorine is bubbled into potassium bromide solution, the solution turns an orange colour. Write a balanced symbol equation to show this reaction.

      (2 marks)

6 A part of the Periodic Table is shown below.

													H									He
Li	Be															B	C	N	O	F		Ne
Na	Mg															Al	Si	P	S		Cl	Ar
K	Ca										Cu	Zn		Ge					Br		Kr
Rb	Sr												Sn					I			Xe

a) Name the Russian scientist whose work led to the development of the modern Periodic Table.

      (1 mark)

b) Using only the elements shown above name

   (i) the **most** reactive alkali metal

      (1 mark)

   (ii) the **least** reactive halogen

      (1 mark)

   (iii) the **least dense** noble gas

      (1 mark)

   (iv) a semi-metal

      (1 mark)

   (v) a metal of valency 3

      (1 mark)

   (vi) **two** different elements which can exist as allotropes

      (2 marks)

7 This question is about the physical and chemical properties of the alkali metals. You may find your Data Leaflet helpful.

a) The table lists some physical properties of alkali metals.

Element	Melting point (°C)	Boiling point (°C)	Density (g/cm³)
Lithium	180	1347	0.53
Sodium	98	883	0.97
Potassium	64	774	0.86
Rubidium	39	688	1.53
Caesium	28	678	1.88

(i) Which of the alkali metals is a liquid at 35°C?

(1 mark)

(ii) How do the boiling points of the alkali metals vary down the group?

(1 mark)

(iii) Name a metal in the table which would sink if placed in water.

(1 mark)

b) Potassium hydroxide, like sodium hydroxide, is a base. It reacts with sulphuric acid to form a solution containing potassium sulphate

(i) Describe how you could obtain crystals of potassium sulphate from potassium sulphate solution.

(2 marks)

(ii) Give the chemical formula of potassium sulphate.

(1 mark)

(iii) Why should a chemist **not** try to make potassium sulphate by reacting potassium with sulphuric acid?

(1 mark)

c) The reaction of the Group I metal rubidium (Rb) with water is given by the word equation

rubidium + water → rubidium + hydrogen
                              hydroxide

(i) Write a balanced symbol equation for this reaction

(2 marks)

(ii) From your knowledge of the reaction of sodium with water predict **three** things you would expect to see, **apart from bubbles of gas**, if a small piece of rubidium was placed in water.

(3 marks)

8 a) Mendeleev was responsible for much of the early development work on the Periodic Table.

(i) Give **two** ways in which Mendeleev's Periodic Table is different from the one we use today.

(2 marks)

(ii) Helium, neon and argon belong to the same group in the Periodic Table. What is the name given to this group and why are its members very unreactive?

(3 marks)

The elements magnesium and calcium are reactive metals with similar chemical properties.

(iii) How can you tell from the Periodic Table that magnesium and calcium are reactive metals with similar chemical properties?

(2 marks)

(iv) Chlorine, bromine and iodine belong to Group VII of the Periodic Table. Which of these elements would you expect to be least reactive and why?

(2 marks)

Chapter 12

Metals and Their Compounds

Learning objectives

By the end of this chapter you will:

➤ Know about the physical properties of metals and about their reactions with oxygen, water (cold water and steam) and acids

➤ Know how a selection of metals can be put into a reactivity series

➤ Know how to work out where an unknown metal would fit into that series

➤ Know how metal ions can be identified by flame tests and by using sodium hydroxide solution or ammonia solution

➤ Know how the ease of extraction of a metal depends on how reactive it is

➤ Know how iron is extracted from its ore

➤ Know about the properties of metal compounds and you will learn of the typical reactions of oxides, hydroxides and carbonates

➤ Know about rust, especially how to prevent it forming on iron or steel

➤ Know about the uses of particular metals and metal compounds

Physical properties

● Metals have high melting and boiling points and so they are solids at room temperature. Mercury, which is the only metal element to be a liquid at room temperature, is the exception.

● Most metals are hard, that is they are not easily scratched, dented or cut. The Group I metals are exceptions.

● Most metals are strong, that is they can easily support a load. Obviously, the strength depends not only on the particular metal but also on the metal's shape and size.

● Most metals are denser than water, i.e. most metals sink when a lump of it is placed in water. Again the Group I metals are exceptions.

● The surface of a metal is, or can be made, shiny. This is also known as having a **lustre** or being lustrous.

● Metals are good electrical conductors.

● Metals are good thermal conductors.

● Metals are **ductile**, they can be drawn into wires.

● Metals are **malleable**, they can be beaten into shape.

123

Chemical properties of metals and the reactivity series

Although many metals take part in the same reactions some metals are more reactive than others.

By observing the behaviour of different metals in the same reactions they can be put into a **reactivity series**.

The metals you are required to know about are:

    aluminium, calcium, copper, iron, magnesium, potassium, sodium and zinc

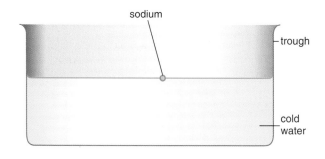

Figure 1 The reaction of sodium with cold water

Reaction with cold water

Only calcium, potassium and sodium react immediately with cold water. A *very* small piece of potassium or sodium should be added with great care to a *large* volume of water. Not only will the metal float but will also move rapidly over the surface. Potassium is more reactive than sodium. With both metals a noise will be heard as the hydrogen evolved burns. If the reaction is vigorous enough potassium will have a lilac flame and sodium a yellow one. The metal hydroxide is produced and this dissolves in the water to give an alkaline solution.

Calcium is much less reactive but care is still needed. The metal sinks but as hydrogen bubbles cling to the piece of metal it may float up. An alkaline solution of calcium hydroxide is produced.

General equation:

metal + cold water → metal hydroxide + hydrogen

Example:

$$2K \ + \ 2H_2O \ \rightarrow \ 2KOH \ + \ H_2$$

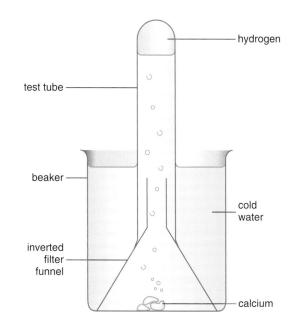

Figure 2 The reaction of calcium with cold water

Reaction with hot water (steam)

Iron, magnesium and zinc will react when heated with steam. The products are the metal oxide and hydrogen. Magnesium is the most reactive and iron the least.

General equation:

metal + steam → metal oxide + hydrogen

Example:

$$Mg \ + \ H_2O \ \rightarrow \ MgO \ + \ H_2$$

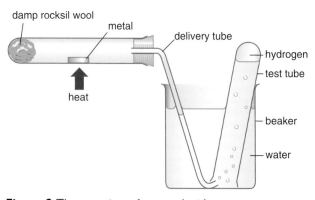

Figure 3 The reaction of a metal with steam

Reaction with dilute acid

Copper, iron, magnesium and zinc can be placed in dilute acid, usually hydrochloric acid. All but copper will react producing the metal salt and giving off hydrogen. As before, magnesium is the most reactive, then zinc and iron is the least reactive.

General equation:

$$metal + acid \rightarrow metal\ salt + hydrogen$$

Example: $$Zn + 2HCl \rightarrow ZnCl_2 + H_2$$

Calcium, potassium and sodium would all, of course, react with dilute acid but these reactions are rarely, if ever, carried out in a school laboratory – they would be much too vigorous!

Displacement reactions

When magnesium metal is placed in a solution of copper sulphate a reaction occurs but if copper metal is placed in a solution of magnesium sulphate there is no reaction. The magnesium can displace the copper from its salt but the copper cannot displace the magnesium.

$$Mg(s) + CuSO_4(aq) \rightarrow MgSO_4(aq) + Cu(s)$$

This is an example of the general rule that a more reactive metal will displace a less reactive metal from a solution of its salt.

It is also an example of a redox reaction.

The magnesium is oxidised as it is losing electrons:

$$Mg \rightarrow Mg^{2+} + 2e^-$$

The copper is reduced because it is gaining electrons:

$$Cu + 2e^- \rightarrow Cu$$

Reactivity of aluminium

Aluminium becomes covered with a very unreactive and firmly attached coating of aluminium oxide that is very difficult to remove. It is this coating that protects the aluminium and allows it to be used where the metal itself would be expected to react, for example overhead electricity cables and saucepans.

In order to place aluminium in the reactivity series metals are added to an aluminium sulphate solution. Magnesium metal reacts with aluminium sulphate solution but zinc metal does not. Aluminium therefore comes after magnesium and before zinc in the series.

Order of reactivity (descending) (and one way to remember it)

Potassium	Poor
Sodium	Scientists
Calcium	Can
Magnesium	Make
Aluminium	A
Zinc	Zoo
Iron	In
Copper	Coleraine

You could try making up your own phrase or sentence. If it is your own work you may find it easier to remember.

Questions

1 In which of the following would a displacement reaction occur?

   a) Calcium and potassium chloride solution
   b) Magnesium and iron nitrate solution
   c) Zinc and copper chloride solution
   d) Iron and sodium sulphate solution
   e) Magnesium and zinc chloride solution

Placing an unfamiliar element in the reactivity series: Examples

You need to be able to use given data to place elements in a reactivity series.

1 Silver does not react with water (hot or cold), it does not react with dilute acid nor does it react with a copper sulphate solution. Where would it fit into the reactivity series given earlier?

Answer: below copper.

It would fit beside copper because of its lack of reaction with water and with dilute acid but because it cannot displace copper from copper sulphate the silver must be less reactive than copper.

2 A new metal has been discovered. It does not react with cold water but it reacts when heated in steam. It was placed in different salt solutions with the following results.

Metal salt soln.	$MgSO_4$	$Al_2(SO_4)_3$	$ZnSO_4$	$FeSO_4$	$CuSO_4$
Observation	no reaction	no reaction	no reaction	reaction	reaction

   Place this new metal in the reactivity series.

   Answer: below zinc and above iron.

The lack of reaction with cold water puts this metal below calcium. The reactions with steam and with dilute acid put the metal above copper. It cannot displace magnesium, aluminium or zinc so it must be below zinc. It can displace iron so it is above iron

3 Four metals: M, N, O and P were added to separate test-tubes of dilute hydrochloric acid. N and O reacted but M and P did not. When metal N was placed in a solution of a salt of metal O there was no reaction. When metal M was placed in a solution of a salt of metal P there was a reaction. Place these metal in descending order of reaction.

   Answer: O, N, M, P.

   N and O are more reactive than M and P because they do react with dilute acid and M and P do not. N is less reactive than O because it cannot displace O from its salt. M is more reactive than P because it can displace P from a solution of its salt.

Identification of metal ions

Flame tests

When compounds containing certain metal ions are placed in a flame they may give a characteristic colour to the flame. These colours can be used in the identification of these metal ions.

Table 1 Flame colours

Metal ion	Flame colour
Ca^{2+}	brick red
Cu^{2+}	green
K^+	lilac
Na^+	yellow

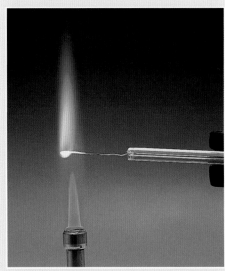

Figure 4 Potassium compounds make the Bunsen flame lilac

Method: take a nichrome wire and dip it into concentrated hydrochloric acid. Place the wire into a blue Bunsen flame. If the wire is clean no colour should appear, if there is some colour repeat the dipping and heating until the wire is clean. Then touch the wire to a solid sample of the compound and put it back into the flame, note the flame colour.

Using sodium hydroxide solution or ammonia solution to precipitate the metal hydroxide

Most metal hydroxides are insoluble and can be formed by precipitation.

$$M^{n+}(aq) + nOH^-(aq) \rightarrow M(OH)_n(s)$$

Several of the hydroxides have a characteristic colour that can be an aid to identification of the metal ion.

Some of the precipitates disappear when excess sodium hydroxide or ammonia solution is added. This is because the metal hydroxide undergoes a further reaction forming a soluble substance.

Metal ion	with NaOH solution			with NH₃ solution		
	precipitate	colour	in excess	precipitate	colour	in excess
K^+	no	—	—	no	—	—
Na^+	no	—	—	no	—	—
Ca^{2+}	yes	white	insoluble	yes	white	insoluble
Mg^{2+}	yes	white	insoluble	yes	white	insoluble
Al^{3+}	yes	white	soluble	yes	white	insoluble
Zn^{2+}	yes	white	soluble	yes	white	soluble
Fe^{2+}	yes	green	insoluble	yes	green	insoluble
Fe^{3+}	yes	brown	insoluble	yes	brown	insoluble
Cu^{2+}	yes	blue	insoluble	yes	blue	soluble

Table 2

Extraction of metals

A few metals such as gold, are found uncombined in nature but nearly all metals are found naturally as compounds. These compounds are found in particular rocks as the metal ores, for example iron is found as iron(III) oxide in haematite.

The metal in the ore is present as ions. When a metal is extracted from its ore the metal ions gain electrons and so the metal is reduced.

$$M^{n+} + ne^- \rightarrow M$$

The more reactive a metal the harder it is to extract it from its ore. Iron and copper are fairly easily extracted from their ores using heat and a chemical reducing agent. Aluminium, however, would require a very strong chemical reducing agent and the industrial process uses electrolysis instead of a chemical reducing agent. (For details about the extraction of aluminium see the chapter on electrolysis.)

Extraction of iron

Iron is extracted from its ore in a blast furnace.

The raw materials are iron ore, limestone and coke, which are fed in to the top of the furnace, and hot air, which is blasted in at the bottom.

There are three main processes taking place in the blast furnace.

1 Formation of the reducing agent

The coke burns to give carbon dioxide; further reaction with more coke produces carbon monoxide which is the reducing agent.

$$C + O_2 \rightarrow CO_2$$
$$C + CO_2 \rightarrow 2CO$$

Overall:
$$2C + O_2 \rightarrow 2CO$$

2 Reduction of the iron oxide to iron

The carbon monoxide reacts with the iron oxide.

$$Fe_2O_3 + 3CO \rightarrow 2Fe + 3CO_2$$

At the temperature of the furnace, up to 2000 °C, the iron is molten. It sinks to the bottom and is run off from there.

3 Removal of the impurity

The main impurity is sand (silicon dioxide).

At the temperature of the furnace the limestone (calcium carbonate) decomposes.

$$CaCO_3 \rightarrow CaO + CO_2$$

The calcium oxide reacts with the silicon dioxide giving calcium silicate (slag).

$$CaO + SiO_2 \rightarrow CaSiO_3$$

The slag is also molten and it runs to the bottom of the furnace where it lies on top of the molten iron. It is tapped off separately.

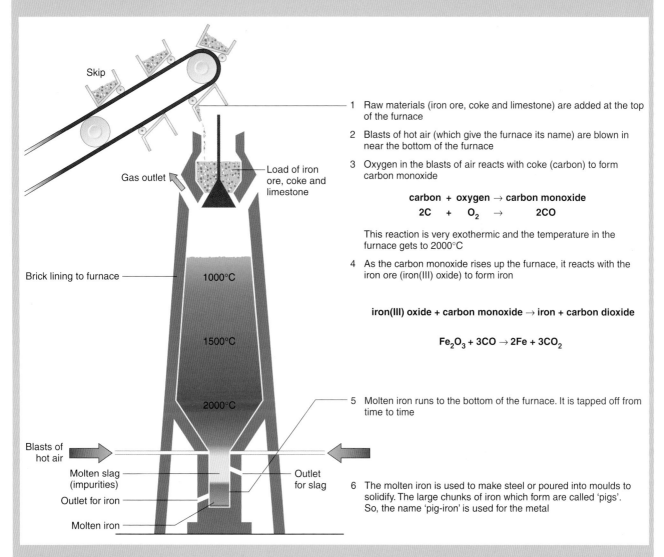

1 Raw materials (iron ore, coke and limestone) are added at the top of the furnace

2 Blasts of hot air (which give the furnace its name) are blown in near the bottom of the furnace

3 Oxygen in the blasts of air reacts with coke (carbon) to form carbon monoxide

carbon + oxygen → carbon monoxide
$$2C \;+\; O_2 \;\rightarrow\; 2CO$$

This reaction is very exothermic and the temperature in the furnace gets to 2000°C

4 As the carbon monoxide rises up the furnace, it reacts with the iron ore (iron(III) oxide) to form iron

iron(III) oxide + carbon monoxide → iron + carbon dioxide

$$Fe_2O_3 + 3CO \rightarrow 2Fe + 3CO_2$$

5 Molten iron runs to the bottom of the furnace. It is tapped off from time to time

6 The molten iron is used to make steel or poured into moulds to solidify. The large chunks of iron which form are called 'pigs'. So, the name 'pig-iron' is used for the metal

Figure 5 Extracting iron from iron ore in a blast furnace

Metal compounds

Physical properties of metal compounds

Nearly all metal compounds are ionic and show the properties of ionic compounds, for example, high melting points, brittleness and the ability to conduct electricity when molten or in solution but not when solid.

Information about the solubility of the metal compounds can be found in the data leaflet provided by CCEA for GCSE Science examinations.

Chemical properties of metal compounds

Oxides and hydroxides of metals

Both oxides and hydroxides of metals are basic and so they react with dilute acids forming salts and water. Some dissolve in water and some react with it.

Compound	with acid	with water
CaO	$CaO + 2HCl \rightarrow CaCl_2 + H_2O$	$CaO + H_2O \rightarrow Ca(OH)_2$
CuO	$CuO + H_2SO_4 \rightarrow CuSO_4 + H_2O$	Insoluble
NaOH	$2NaOH + H_2SO_4 \rightarrow 2NaCl + 2H_2O$	Soluble
$Ca(OH)_2$	$Ca(OH)_2 + 2HCl \rightarrow CaCl_2 + 2H_2O$	Sparingly soluble

Table 3

Sodium (or potassium) hydroxide solution reacts with carbon dioxide and can be used to remove carbon dioxide from the air. This can be used by biologists in experiments on photosynthesis when they need a carbon dioxide-free atmosphere

$$2NaOH(aq) + CO_2(g) \rightarrow Na_2CO_3\ (aq) + H_2O(l)$$

This reaction is responsible for the white powdery crust that is sometimes seen around the neck of bottles containing sodium hydroxide solution. When the solution is poured out a thin film is left on the bottle, carbon dioxide from the air reacts with the sodium hydroxide and then the water evaporates leaving solid sodium carbonate.

Calcium hydroxide solution, limewater, undergoes the same reaction but the carbonate formed is insoluble so it is precipitated out of solution. This is why limewater goes milky or cloudy when carbon dioxide is bubbled through it.

$$Ca(OH)_2(aq) + CO_2(g) \rightarrow CaCO_3(s) + H_2O(l)$$

Sodium hydroxide reacts with ammonium salts to form ammonia.

$$NaOH(aq) + NH_4Cl(aq) \rightarrow NaCl(aq) + NH_3(g) + H_2O(l)$$

The test for the presence of the ammonium cation involves heating the suspected ammonium salt with sodium hydroxide solution and testing for an alkaline gas (the ammonia) being evolved.

Rust, an economically important oxide

Iron, usually as steel, has many uses because of its strength and relatively low cost. Unfortunately iron rusts. Knowledge about how rust is formed and how to prevent or reduce the rusting process is of great economic importance.

Formation of rust

Iron reacts with oxygen in the presence of water to form rust.

IRON + OXYGEN + WATER → RUST

The full chemical name of rust is hydrated iron(III) oxide, $Fe_2O_3.2H_2O$

$$4Fe + 3O_2 + 4H_2O \rightarrow 2Fe_2O_3.2H_2O$$

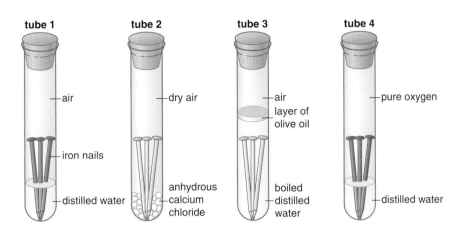

Figure 6 Investigating whether water and oxygen are involved in rusting

In the above experiment the iron nails in tubes 1 and 4 rust.

Questions

2   What is the reason for having calcium chloride in tube 2?

3   Why do the iron nails in tube 2 not rust?

4   In tube 3 what was removed from the water when it was boiled?

5   Why do the nails in tube 3 not rust?

6   What conditions does this experiment show are needed for iron to rust?

Prevention

Most methods of rust prevention involve covering the iron with a material that acts as a barrier stopping air (oxygen) and moisture from coming in contact with the iron. Common methods are painting, oiling or greasing, tin-plating and galvanising (covering with a layer of zinc).

Sacrificial protection involves the use of another metal that reacts before the iron does. This other metal, therefore, must be higher in the reactivity series than iron is.

Ships' hulls and marine oil rigs are made of steel, yet they are used in water where rusting is going to be a serious problem. Attaching bars of zinc or magnesium to hulls or rig legs greatly reduces the rust formation. In each case the more reactive metal corrodes first. It is 'sacrificed' to protect the iron.

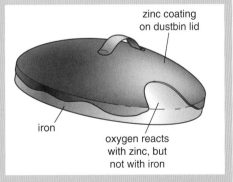

Figure 7 On a galvanized dust bin, zinc reacts before the iron, even where the zinc coating is scratched or broken

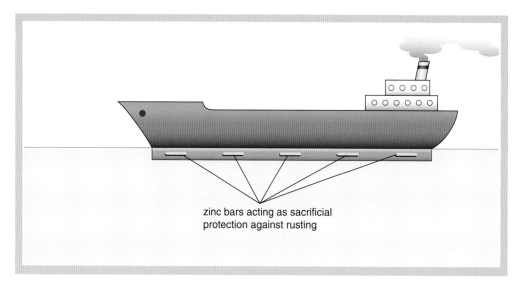

Figure 8 Sacrificial protection of a ship's hull

zinc bars acting as sacrificial
protection against rusting

Carbonates

Most metal carbonates undergo thermal decomposition, i.e. they break down on heating.

metal carbonate → metal oxide + carbon dioxide

Example: $CuCO_3$ →   CuO   + CO_2
        green   black

The decomposition of calcium carbonate (limestone) is important in industry e.g. in the extraction of iron and in the production of quicklime and of slaked lime.

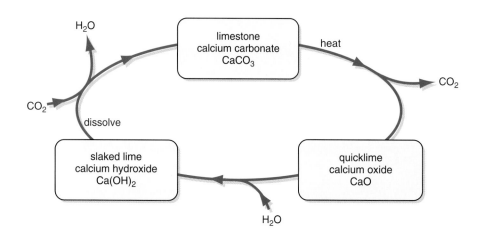

Figure 9 The lime cycle

The less reactive the metal the less stable is the carbonate. In other words, the less reactive the metal the easier it is to decompose the carbonate by heating. Heating cannot decompose the carbonates of sodium and potassium because they are too stable.

All metal hydrogencarbonates are decomposed by heat.

metal hydrogencarbonate → metal carbonate + carbon dioxide + water

Example: $2NaHCO_3$ → Na_2CO_3 + CO_2 + H_2O

Carbonates and hydrogencarbonates react with acid to form salt, carbon dioxide and water. (See also chapter on acids and bases and the preparation of carbon dioxide.)

$$Na_2CO_3 + 2HCl \rightarrow 2NaCl + CO_2 + H_2O$$
$$CuCO_3 + H_2SO_4 \rightarrow Cu\,SO_4 + CO_2 + H_2O$$

Uses of metals

● Magnesium is combined with other metals to provide a high strength but low density alloy for building aircraft bodies. Because magnesium burns with a brilliant white flame it is used in flares.

● Aluminium is a good conductor of electricity and has a relatively low density so it is used to make overhead electric cables. It is also a good thermal conductor, and because of the oxide coating, it does not react with hot water and so it can be used to make saucepans. Its low density means it is suitable for lightweight alloys.

● Zinc is used to cover iron and prevent it from rusting, this process is known as galvanising. Zinc is alloyed with copper to form brass.

● Iron, as steel, has very many uses, for example in the construction of buildings, ships' hulls, bridges, tools, car bodies, cookers and fridges. It is strong and cheap. Iron is essential in our diet because it is needed in the formation of haemoglobin the complex molecule found in red blood cells, that carries oxygen around our bodies.

● Copper is a good conductor of electricity and is used in electrical wiring. As it does not react with water even when hot it is used to make water pipes and hot water cylinders. Copper is also used to make alloys that are suitable to make coins.

● Lead is soft, very malleable and does not react with water which makes it very suitable to be used where flexibility is needed in roofing where it is necessary to go around corners and where tiles or slates would not bend. It is also used in car batteries, in solder and, of far less consequence today, to make the anti-knock compound that was present in leaded petrol.

● Calcium is needed in our diet to help make strong bones and teeth.

Uses of metal compounds

● Sodium chloride mixed with grit is spread on the roads in winter. The sodium chloride lowers the melting point of water making ice form at a lower temperature and the grit provides grip if ice does form.

● Sodium hydrogen carbonate is also known as baking soda and is used in cookery as a raising agent. The carbon dioxide produced by its reaction with acid is trapped in the mixture and makes it rise. Many indigestion remedies contain sodium hydrogen carbonate. It is suitable as an antacid because it reacts to neutralise stomach acid.

● Calcium carbonate (limestone) has many uses, for example cement manufacture, in blackboard chalk and for slag removal in the blast furnace.

● Calcium hydroxide (slaked lime) is used in agriculture to reduce the acidity of soil.

● Calcium sulphate is used in Plaster of Paris, in blackboard chalk and in the mixture used to mark white lines on the roads.

● Aluminium hydroxide has been used as an antacid but is much less popular now because of concerns about the connection between aluminium and Alzheimer's disease.

Websites

www.gcsechemistry.com/ukop.htm

www.s-cool.co.uk

www.schoolscience.co.uk

Exam questions

1 Some students compared the reactions of four metals by looking at how each metal reacts in turn with the sulphate solutions of the other three metals. The results are shown in the table below.

metal \ solution	cobalt(II) sulphate	iron(II) sulphate	copper(II) sulphate	magnesium sulphate
cobalt		no reaction	reacts quite slowly	no reaction
iron	reacts very slowly		reacts quite quickly	no reaction
copper	no reaction	no reaction		no reaction
magnesium	reacts quickly	reacts quickly	reacts quickly	

a) Apart from reacting quickly, what else would you observe during the reaction of magnesium with copper(II) sulphate?
(1 mark)

b) Write a **word** equation for the reaction of magnesium with copper(II) sulphate.
(1 mark)

c) List the four metals cobalt, iron, copper and magnesium in order of reactivity, putting the most reactive metal first.
(2 marks)

d) Which of these four metals would you expect to react most quickly with oxygen? Give a reason for your answer.
(1 mark)

2 The reaction of magnesium with cold water is very slow. However, it does react well if heated with steam.

a) Draw a labelled diagram of the apparatus used to carry out this reaction and to collect any gaseous product(s) formed.
(6 marks)

b) List **one** safety precaution you would take
(i) during the reaction   *(1 mark)*
(ii) after the reaction has just finished.
(1 mark)

c) Describe how the appearance of the solid changes during the reaction.
(2 marks)

d) Write a balanced, symbol equation for the reaction.
(2 marks)

e) Is the magnesium oxidised or reduced during the reaction? Explain your answer.
(2 marks)

f) This reaction is classified as **exothermic**. What do you understand by the term 'exothermic'?

(2 marks)

g) One important use of magnesium is the prevention of rusting by sacrificial protection. Explain briefly how this process works.

(4 marks)

h) Other than its use in sacrificial protection, write down **two** uses of magnesium.

(2 marks)

3 The list below contains some metals in the reactivity series.

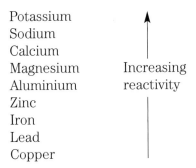

Potassium
Sodium
Calcium
Magnesium   Increasing
Aluminium   reactivity
Zinc
Iron
Lead
Copper

a) (i) Which metal in the list is **least likely** to react with dilute hydrochloric acid?

(1 mark)

(ii) Which reaction, A, B or C, given below, could **not** take place? Give a reason for your answer.

A   calcium + aluminium oxide → calcium oxide + aluminium

B   aluminium + iron oxide → aluminium oxide + iron

C   iron oxide + calcium oxide → iron + calcium

(3 marks)

b) (i) Which **two** words below can be used to describe most metal oxides at room temperature (20°C).

acids   bases   solids   liquids   gases

(2 marks)

c) Lead is a soft dense metal with a low melting point. It is used as 'flashing' on roofs, and is also used in solder. Water pipes were made from lead until more suitable materials became available.

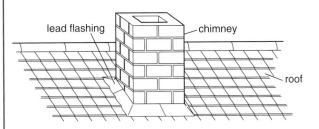

(i) Which metal is now used for making water pipes?

(1 mark)

(ii) Name a material, other than a metal, which can be used for making water pipes and give two advantages of this material.

(3 marks)

(iii) Why is **lead** used as 'flashing' on roofs?

(1 mark)

(iv) What property of lead makes it useful for solder?

(1 mark)

d) Use the Periodic Table in your Data Leaflet to identify:

(i) A metal in the same period as sodium.

(1 mark)

(ii) A metal which has **two** electrons in its outer shell.

(1 mark)

(iii) A metal which has 13 protons.

(1 mark)

4 a) Iron is the cheapest and most important metal available to man. It is produced in large quantities from its ore in the blast furnace.
Name the ore from which iron is produced.

(1 mark)

b) Describe the production of iron from its ore in the blast furnace. Your answer should include the names of the raw materials used, how the silicon dioxide impurities are removed and balanced symbol equations for the reactions that occur.

(14 marks)

c) Although iron is a very useful metal in the service of man its uses are sometimes limited because it rusts.

(i) What conditions are needed to make iron rust?

(2 marks)

(ii) What is the correct chemical name for rust?

(2 marks)

(iii) Is the formation of rust an oxidation or reduction reaction? Explain your answer.

(2 marks)

d) Underground steel pipes have been protected from corrosion by connecting easily replaceable magnesium rods along them at specific points. Explain as fully as you can why this method of preventing rusting works.

(2 marks)

5   An experiment was set up to find out what causes iron nails to rust.

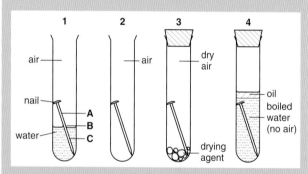

Rusting was fastest in test-tube 1. Rusting occurred slowly in test-tube 2 but there was no rusting in the other two test-tubes.

a) At what point, **A**, **B** or **C**, would you expect rust to form first in test-tube 1? Give your reasons for your answer.

(1 mark)

b) Painting can protect iron railings from rusting. Give **one** advantage and **one** disadvantage of using paint to protect iron from rusting.

(2 marks)

6   a) Limestone has many uses in industry. It is often heated in large lime kilns to produce quicklime (calcium oxide).

(i) What is the chemical name for limestone?

(1 mark)

(ii) Write a **word** equation to show what happens when limestone is heated.

(1 mark)

(iii) What term is used to describe the type of reaction that has occurred when limestone is heated?

(2 marks)

(iv) Give **one** use for limestone (apart from making quicklime).

(1 mark)

b) (i) Calcium oxide can be converted into calcium hydroxide. Write a **word** equation to show how this is done.

(1 mark)

(ii) What would you observe if this reaction was carried out on a watch-glass?

(3 marks)

c) Calcium hydroxide is sparingly soluble into water. A mixture of water and calcium hydroxide is stirred before it is filtered.

(i) What is the common name of the filtrate?

(1 mark)

(ii) The filtrate slowly begins to turn cloudy if it is allowed to remain exposed to the air. What substance in air do you think is responsible for this?

(1 mark)

(iii) Write a **word** equation for the reaction that has occurred in part (ii) above.

(1 mark)

Chapter 13

Non-metals and Their Compounds

Learning objectives

By the end of this chapter you will know:

➤ How to test for hydrogen, oxygen and carbon dioxide

➤ How to test for ammonia and hydrogen chloride

➤ The meaning of the word 'diatomic'

➤ How to prepare hydrogen and oxygen, in the laboratory.

➤ How to prepare chlorine and hydrogen chloride in the laboratory

➤ About the physical properties of hydrogen, carbon dioxide, nitrogen, oxygen, sulphur chlorine

➤ About the physical properties of the other halogens

➤ About some of the reactions of hydrogen, carbon, carbon dioxide, oxygen, sulphur, sulphur dioxide and chlorine

➤ About some of the reactions of ammonia

➤ That the noble gases are chemically inert and that nitrogen is very unreactive

➤ the industrial production of ammonia

➤ About the industrial production of sulphuric acid and nitric acid

Hydrogen

Hydrogen is the most abundant element in the Universe.

Preparation

Acid plus metal forms salt plus hydrogen and this is the method used in the laboratory preparation of hydrogen. The metal is usually zinc or magnesium and the acid is hydrochloric.

$$Zn + 2HCl \rightarrow ZnCl_2 + H_2$$

Physical properties

Hydrogen is a colourless, odourless, neutral gas. It is the lightest element, in other words it has the lowest density. It is insoluble in water.

It consists of simple molecules containing two covalently bonded hydrogen atoms. Any molecule with only two atoms in it is described as **diatomic**.

Test

When a lighted splint is place in a test tube of hydrogen there is a squeaky pop. This is a mini-explosion because the hydrogen burns very quickly forming water

$$2H_2 + O_2 \rightarrow H_2O$$

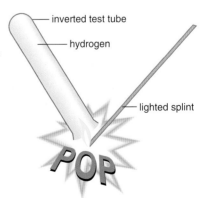

inverted test tube

hydrogen

lighted splint

POP

Figure 1 The 'pop' test for hydrogen

Reactions

- With oxygen: hydrogen reacts with oxygen to give water (see test).
- As a reducing agent: when hydrogen is heated with certain metal oxides the metal is reduced. However, only the metals below iron in the reactivity series can be reduced by hydrogen.

$$CuO + H_2 \quad \rightarrow \quad Cu \quad + \quad H_2O$$
$$\text{black} \qquad\qquad \text{red/brown}$$

- With nitrogen: hydrogen can be made to react directly with nitrogen to form ammonia (see the Haber–Bosch process page 141).

Questions

1   What apparatus would you need and how would you set it up to demonstrate, in the school laboratory, the reduction of copper oxide by hydrogen?

2   Why can the reaction between hydrogen and nitrogen not be carried out in a school laboratory?

Carbon

All life on Earth is based on the element carbon.

Physical properties

Diamond and graphite are the two main allotropes of carbon, they have giant molecular structures. (See Chapter 2 Chemical Bonding and the Properties of Materials.)

In the 1980s another allotrope was discovered in which the carbon atoms are joined in polygons of five or six atoms and then these circles are joined to give large spherical or tubular molecules. These structures are called fullerenes and they got their name from their likeness to the geodesic dome structures designed by R. Buckminster Fuller. The first one to be identified was called Buckminster Fullerene (nicknamed 'buckyball'). It has 60 carbon atoms arranged in 32 polygons and looks like a football.

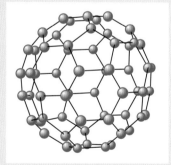

Figure 2 The structure of C_{60}

Figure 3 The structure of bucky tubes

Combustion

Carbon burns to give carbon dioxide:

$$C + O_2 \rightarrow CO_2$$

This is described as complete combustion.

If there is a limited amount of oxygen present then the combustion is incomplete and carbon monoxide is formed:

$$2C + O_2 \rightarrow 2CO$$
$$\text{or} \quad C + CO_2 \rightarrow 2CO$$

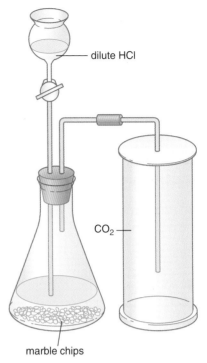

dilute HCl

CO_2

marble chips

Figure 4 Making carbon dioxide in the laboratory. **Wear eye protection** if you are doing this

Questions

3 Why it is very important that whenever carbon or carbon containing compounds are burned there is a good supply of air?

Carbon dioxide

Preparation

The usual method of preparing carbon dioxide in the laboratory is to react hydrochloric acid with calcium carbonate in the form of marble chips.

$$2HCl + CaCO_3 \rightarrow CaCl_2 + CO_2 + H_2O$$

It is possible to use other acids and carbonates:

$$2HNO_3 + Na_2CO_3 \rightarrow 2NaNO_3 + CO_2 + H_2O$$
$$H_2SO_4 + MgCO_3 \rightarrow MgSO_4 + CO_2 + H_2O$$

Questions

4 Why would the combination of sulphuric acid and calcium carbonate not be a suitable method of preparing carbon dioxide?

Physical properties

Carbon dioxide consists of simple molecules with one carbon atom covalently bonded to two oxygen atoms. Its physical properties are shown in the diagram.

Test

When carbon dioxide is bubbled through limewater the limewater goes cloudy (or milky).

$$CO_2(g) + Ca(OH)_2(aq) \rightarrow CaCO_3(s) + H_2O(l)$$

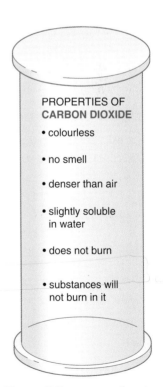

PROPERTIES OF CARBON DIOXIDE

• colourless

• no smell

• denser than air

• slightly soluble in water

• does not burn

• substances will not burn in it

Figure 5 Properties of carbon dioxide

Questions

5   Which product causes the cloudiness when carbon dioxide is bubbled through limewater?

6   What type of reaction is this?

Reactions

- With water: carbon dioxide dissolves in water and reacts to give the weak acid carbonic acid.

$$CO_2 + H_2O \rightarrow H_2CO_3$$

- With sodium hydroxide: when carbon dioxide is bubbled into sodium hydroxide solution sodium carbonate and water are formed.

$$CO_2(g) + 2NaOH(aq) \rightarrow Na_2CO_3(aq) + H_2O(l)$$

As with the reaction between carbon dioxide and calcium hydroxide solution (limewater) this is an acid–base reaction with the carbon dioxide acting as an acid. Unlike the reaction with limewater the mixture does not go cloudy. Sodium carbonate is soluble and is not precipitated.

- With burning magnesium: when a burning piece of magnesium ribbon is lowered into a gas jar of carbon dioxide it continues to burn. The magnesium is able to obtain oxygen from the carbon dioxide.

$$2Mg + CO_2 \rightarrow 2MgO + C$$

A white powder (magnesium oxide) with black specks (carbon) is left on the sides of the gas jar.

Carbon dioxide pollution

Although carbon dioxide is essential to life on earth, (it is required for photosynthesis), it is also a greenhouse gas. (See Chapter 14 Organic Chemistry.)

Nitrogen

Nitrogen is the most abundant element in the Earth's atmosphere.

Physical properties and Lack of reactivity

Nitrogen is a colourless, odourless, neutral gas. It is only very sparingly soluble in water.

It consists of simple diatomic molecules in which the atoms are held together by a strong triple covalent bond. This bond is very hard to break and so nitrogen takes part in very few reactions.

Production of ammonia, the Haber–Bosch Process

Nitrogen is very important for good growth of plants and fertilisers containing nitrogen are needed to help grow our food. Man-made fertilisers used to include nitrate from natural deposits but until the 1930s no use was made of nitrogen from the air to provide the ammonia and nitric acid needed to make the fertilisers. A German chemist, Fritz Haber, discovered how to get nitrogen to react directly with hydrogen to form ammonia. Bosch was the chemical engineer who developed Haber's work into an industrial process.

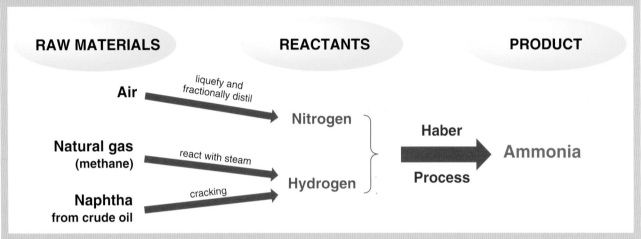

Figure 6 A flow scheme for the Haber process

Nitrogen, obtained from the air, and hydrogen, from natural gas or naphtha, are combined together to produce ammonia.

$$N_2 + 3H_2 \rightarrow 2NH_3$$

The process has to be carried out at high temperature, (450°C), and high pressure, (250 atm) and in the presence of iron as a catalyst.

Only about a quarter of the nitrogen and hydrogen are turned into ammonia and so, to prevent waste, the unreacted gases are recycled. (See also Chapter 16 Rates of Reaction.)

Ammonia

Physical properties

See Figure 7.

Test

Ammonia reacts with hydrogen chloride to give white fumes of ammonium chloride.

$$NH_3(g) + HCl(g) \rightarrow NH_4Cl(s)$$

In this reaction ammonia is acting as a base but unlike other acid-base reactions the only product is the salt and no water is formed.

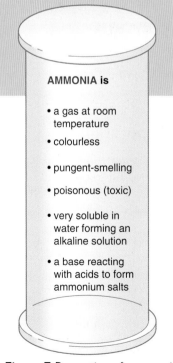

AMMONIA is

- a gas at room temperature
- colourless
- pungent-smelling
- poisonous (toxic)
- very soluble in water forming an alkaline solution
- a base reacting with acids to form ammonium salts

Figure 7 Properties of ammonia

Nitric acid

Production

The raw materials required are ammonia, air (oxygen) and water to form the acid and platinum as the catalyst.

Stage 1: ammonia is reacted, at 900°C, with ten times its volume of air (to make sure there is excess oxygen), in the presence of the platinum to form nitrogen monoxide and water.

$$4NH_3(g) + 5O_2(g) \rightarrow 4NO(g) + 6H_2O(g)$$

Stage 2: the nitrogen monoxide is oxidised to nitrogen dioxide

$$2NO(g) + O_2(g) \rightarrow 2NO_2(g)$$

Stage 3: the nitrogen dioxide reacts with more oxygen and water to form nitric acid.

$$4NO_2(g) + O_2(g) + 2H_2O(l) \rightarrow 4HNO_3(aq)$$

Reactions of dilute nitric acid

Metal oxides, hydroxides and carbonates react with dilute nitric acid as would be expected.

$$CuO + 2HNO_3 \rightarrow Cu(NO_3)_2 + H_2O$$
$$KOH + 2HNO_3 \rightarrow KNO_3 + H_2O$$
$$CaCO_3 + 2HNO_3 \rightarrow Ca(NO_3)_2 + CO_2 + H_2O$$

Oxygen

Oxygen is the most abundant element in the Earth's crust.

Preparation

Hydrogen peroxide decomposes to form water and oxygen. The rate of decomposition is increased by heat and light. In the laboratory preparation of oxygen manganese dioxide is used as a catalyst to speed up the decomposition.

$$2H_2O_2 \rightarrow 2H_2O + O_2$$

Questions

7   Why is hydrogen peroxide is usually kept in a dark glass bottle in a cupboard?

Physical properties

Oxygen consists of simple diatomic molecules. Its properties are listed in Figure 8.

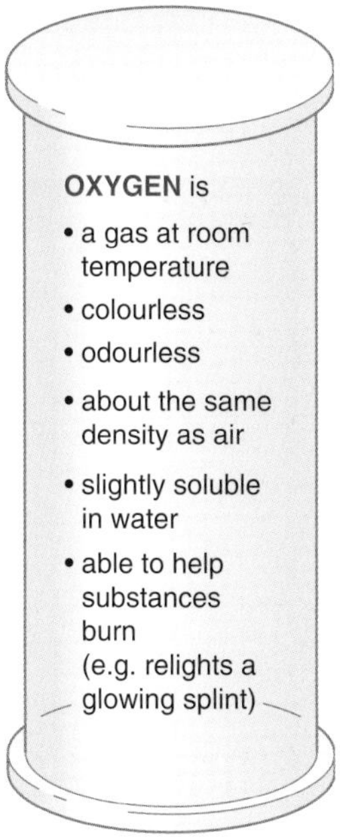

Test

When a glowing splint is placed in a test tube of oxygen the splint relights.

Combustion and respiration

Combustion is the reaction of a fuel with oxygen that produces at least one oxide and releases energy. Burning is an example of combustion.

> fuel + oxygen → carbon dioxide + water + heat

Combustion will not occur without oxygen. To stop something burning it is therefore necessary to prevent oxygen from reaching it.

Respiration is a special form of combustion where the fuel is glucose.

$$\text{glucose} + \text{oxygen} \rightarrow \text{carbon dioxide} + \text{water}$$
$$C_6H_{12}O_6 + 6O_2 \rightarrow 6CO_2 + 6H_2O$$

The process, in the cells of living organisms, is carefully controlled so that the energy is released slowly and in small amounts.

Formation of oxides

Oxygen reacts directly with many metals and non-metals to form their oxides.

OXYGEN is
- a gas at room temperature
- colourless
- odourless
- about the same density as air
- slightly soluble in water
- able to help substances burn (e.g. relights a glowing splint)

Figure 8 Properties of oxygen

Table 1 Comparing the reactions of some elements with oxygen

Element	Reaction with oxygen	Product	Add water to product, then add universal indicator
Sodium	Bright yellow flame White smoke and powder	White solid (sodium oxide)	Dissolves pH = 11 alkaline
Magnesium	Dazzling white flame White clouds and powder	White solid (magnesium oxide)	Dissolves slightly pH = 8 alkaline
Iron	Glows red hot Burns with sparks	Black-brown solid (iron oxide)	Insoluble
Copper	Does not burn however the surface turns black	Black solid (copper oxide)	Insoluble
Carbon	Glows red hot Reacts slowly	Colourless gas (carbon dioxide)	Dissolves pH = 5 acidic
Sulphur	Burns readily with a blue flame	Colourless gas (sulphur dioxide)	Dissolves pH = 3 acidic

8 Look at the table and answer the questions that follow it.

Gas	With lighted splint	With glowing splint	With lime water
A	Splint went out	Splint went out	Lime water cloudy
B	Squeaky pop	Faint pop	No change
C	Splint went out	Splint went out	No change

a) Which gas is hydrogen? Explain your answer.

b) Which gas could be nitrogen? Explain your answer.

c) Name gas A. Explain your answer.

Sulphur

Allotropes

There are three allotropes of sulphur, **rhombic**, **monoclinic** and **amorphous**. Rhombic sulphur exists as bipyramidal crystals and monoclinc sulphur has needle-shaped crystals. Amorphous sulphur is not crystalline and has no definite shape.

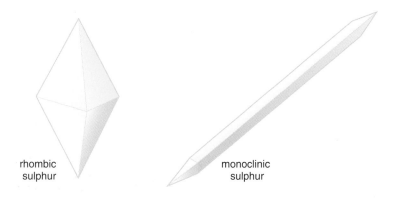

rhombic sulphur

monoclinic sulphur

Figure 9 Two allotropes of sulphur

Physical properties

Sulphur is a yellow solid with a low melting point and is insoluble in water.

Reactions

● Combustion:
Sulphur burns with a blue flame to give sulphur dioxide.

$$S + O_2 \rightarrow SO_2$$

● Reaction with iron:
When iron filings are heated with powdered sulphur they combine to form iron(II) sulphide

$$Fe + S \rightarrow FeS$$

Pollution

Sulphur dioxide reacts with water to form sulphurous acid.

$$SO_{2\ +}\ H_2O \rightarrow H_2SO_3$$

Unfortunately most fossil fuels contain some sulphur and so when they are burned sulphur dioxide is released into the air. Once there it reacts with the water vapour also present and the sulphurous acid formed is returned to the ground in the rain. This acid rain harms plants, damages buildings and corrodes metals.

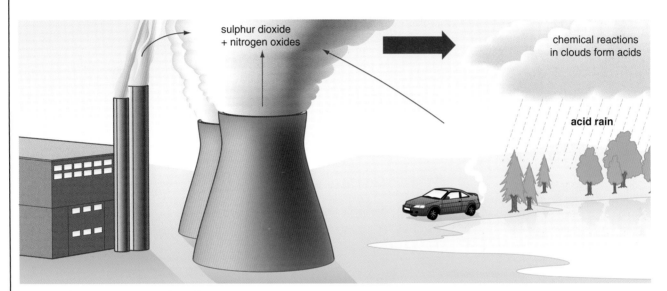

sulphur dioxide + nitrogen oxides

chemical reactions in clouds form acids

acid rain

Figure 10 The formation of acid rain

Sulphuric acid

Reactions and use as a drying agent

The dilute acid undergoes typical reactions with metals and with metal oxides, hydroxides and carbonates.

$$
\begin{aligned}
H_2SO_4 +\ & Mg\ \rightarrow MgSO_4 +\ H_2 \\
H_2SO_4 +\ & ZnO\ \rightarrow ZnSO_4 +\ H_2O \\
H_2SO_4 +\ & 2NaOH \rightarrow Na_2SO_4 +\ 2H_2O \\
H_2SO_4 +\ & CuCO_3 \rightarrow Na_2SO_4 +\ CO_2\ +\ H_2O
\end{aligned}
$$

The concentrated acid has a very strong affinity for water and therefore can be used as a drying agent.

If a drop of concentrated acid is placed on blue, hydrated copper sulphate crystals they can be seen to go pale and then white as the water of crystallisation is removed.

$$CuSO_4.5H_2O \rightarrow\ CuSO_4 +\ 5H_2O$$
blue   white

Even more dramatic is the reaction when concentrated sulphuric acid is mixed with sucrose (ordinary household sugar). The elements of water are removed from the sugar leaving a black solid that is sugar charcoal, a form of carbon. The process is exothermic and as the sugar heats up there is a pleasant toffee-like smell.

$$C_6H_{22}O_{11} \rightarrow \underset{\text{black}}{6C} + 11H_2O$$
$$\underset{\text{white}}{}$$

Production

Sulphuric acid is produced by the Contact Process. It is of great economic importance because the acid has very many uses.

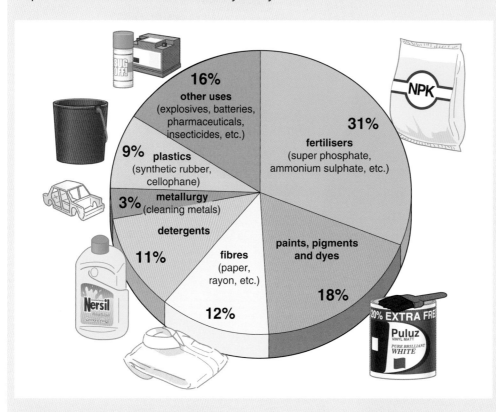

Figure 11 The uses of sulphuric acid

Stage 1: Forming sulphur dioxide.
Sulphur is mixed with air and heated so that it burns to give sulphur dioxide

$$S + O_2 \rightarrow SO_2$$

Stage 2: Oxidation of sulphur dioxide.
The sulphur dioxide is mixed with more air, heated to 450°C and passed over a catalyst of vanadium pentoxide.

$$2SO_2 + O_2 \rightarrow 2SO_3$$

Stage 3: Formation of oleum.
Sulphur trioxide reacts directly with water to form sulphuric acid but it is a very exothermic process and is not used in industry. Instead the sulphur trioxide is absorbed in concentrated sulphuric acid to form oleum.

$$SO_3 + H_2SO_4 \rightarrow H_2S_2O_7$$

Stage 4: Dilution of the oleum.
When the oleum is mixed with water sulphuric acid is formed.

$$H_2S_2O_7 + H_2O \rightarrow 2H_2SO_4$$

This gives 2 moles of the acid – a new one and the one that was used to form the oleum.

Handling

As sulphuric is a strong, corrosive acid it must be handled with great care. The mixing of the concentrated acid with water is a highly exothermic process. The acid must always be added slowly to water to ensure that the water keeps the mixture cool.

Noble gases

Chemical inertness

Helium, neon and argon are chemically very inert and have not been shown to form compounds. Compounds of xenon have been made but extreme conditions are required for their formation.

Questions

9   Why do helium, neon and argon not form chemical bonds?

Halogens (Group VII)

Physical properties of the first four halogens

Halogen	Colour	State at room temperature
Fluorine, F_2	pale yellow	gas
Chlorine, Cl_2	green	gas
Bromine, Br_2	brown	liquid
Iodine, I_2	grey (solid), purple (vapour)	solid

Table 2

When heated, iodine sublimes.

Chlorine

Preparation

Hydrogen chloride can be oxidised to yield chlorine gas. To prepare chlorine in the laboratory concentrated hydrochloric acid is added to potassium manganate(VII).

$$2HCl + [O] \rightarrow H_2O + Cl_2$$

[O] means that atoms of oxygen are being produced by a process which has not been included in the equation.

Properties and poisonous nature

Chlorine reacts with water giving an acidic solution containing hydrochloric acid and chloric(I) acid, (hypochlorous acid).

$$Cl_2 + H_2O \rightarrow HCl + HClO$$

The chloric(I) acid is a bleach and that is why when chlorine gas is tested with blue litmus paper the paper turns first red and then the colour fades. The bleaching of the paper is used as a test for chlorine.

Great care is required when working with chlorine because it is very poisonous. Any reactions attempted in the school laboratory should be carried out in a fume cupboard and very small amounts should be used.

Reactions

● Displacement reactions with bromides and iodides
Reactivity decreases with increase in atomic number in Group 7 thus chlorine is more reactive than bromine and iodine and displaces them from their salts.

$$Cl_2(g) + 2NaBr(aq) \rightarrow 2NaCl(aq) + Br_2(aq)$$

The bromine solution is pale yellow. Adding hexane and shaking the mixture extracts the bromine from the water into the hexane. When the two layers separate the brown colour of bromine can be seen in the hexane layer.

$$Cl_2(g) + 2KI(aq) \rightarrow 2KCl(aq) + I_2(aq)$$

The iodine gives the mixture a light brown colour but when extracted by hexane it turns the hexane layer pale purple.

properties of
CHLORINE

• pale green gas

• choking smell

• poisonous (toxic)

• denser than air

• dissolves in and reacts with water

• bleaches dyes and indicators

• reacts vigorously with most metals

Figure 12 Properties of chlorine

Questions

10 In which of the following would there be a displacement reaction?

a) $I_2(g)$ and KBr(aq)   b) $F_2(g)$ and KBr(aq)

c) $Br_2(g)$ and NaCl(aq)   d) $Cl_2(g)$ and NaF(aq)

e) $Br_2(g)$ and LiBr(aq)

● Direct reaction with elements
Chlorine is a very reactive element and combines directly with many metal and non-metal elements.

Examples: $Cl_2(g) + H_2(g) \rightarrow 2HCl(g)$

$3Cl_2(g) + 2Fe(s) \rightarrow 2FeCl_3(s)$

The chlorine is a strong oxidising agent so it forms the iron(III) ion which is the most oxidised form of iron.

● Reaction with cold sodium chloride solution

$$Cl_2 + 2NaOH \rightarrow NaCl + NaClO + H_2O$$

The sodium chlorate(I) (hypochlorite) is a bleach.

● Reaction with iron(II) ion

When chlorine gas is bubbled through a solution containing the iron(II) ion the solution changes from very pale green to a yellow-brown colour, this is because the chlorine is oxidising the iron(II) to iron(III).

$$Cl_2 + 2Fe^{2+} \rightarrow 2Cl^- + 2Fe^{3+}$$

The chlorine oxidises the iron and is itself reduced.

$$Fe^{2+} \rightarrow 2Fe^{3+} + e^- \quad \text{oxidation (loss of } e^-\text{)}$$
$$Cl_2 + e^- \rightarrow 2Cl^- \quad \text{reduction (gain of } e^-\text{)}$$

Hydrogen chloride

Preparation

The reaction between concentrated sulphuric acid and sodium chloride produces sodium hydrogen sulphate and hydrogen chloride.

$$H_2SO_4 + NaCl \rightarrow NaHSO_4 + HCl$$

Test

Hydrogen chloride reacts with ammonia to give white fumes of ammonium chloride.

$$NH_3(g) + HCl(g) \rightarrow NH_4Cl(s)$$

(As you can see this is the same reaction as the test for ammonia, the purpose of the test depends on which is the unknown gas.)

Reaction with water

Hydrogen chloride gas consists of covalently bonded molecules but when the gas dissolves in water the covalent bond breaks unevenly forming a hydrogen cation and a chloride anion. This solution of hydrogen chloride is hydrochloric acid.

$$HCl(g) + aq \rightarrow H^+(aq) + Cl^-(aq)$$

Questions

11 How does the bonding of hydrogen chloride gas differ from the bonding in hydrochloric acid? Diagrams can be used to aid your explanation.

Hydrochloric acid

The dilute acid reacts with metals, metal oxides, hydroxides and carbonates in typical acid manner.

$$2HCl + Fe \rightarrow FeCl_2 + H_2$$
$$2HCl + CaO \rightarrow CaCl_2 + H_2O$$
$$2HCl + Mg(OH)_2 \rightarrow MgCl_2 + 2H_2O$$
$$2HCl + ZnCO_3 \rightarrow ZnCl_2 + CO_2 + 2H_2O$$

Uses of non-metals

- **Hydrogen** is used to fill meteorological balloons and as a fuel in rocket engines. There is much interest in its possible use as a clean fuel because when it burns the only product is water.
- **Nitrogen** is a liquid below −196°C and is used as a coolant. It can also be used as an inert atmosphere in the packing of certain foods.
- **Sulphur** is added to rubber, a process known as vulcanising, in order to make the rubber harder at high temperatures and less brittle at low temperatures. It is used by gardeners as a fungicide.
- **Oxygen** is supplied to patients who have breathing problems. It is used in welding so that the welding torch works at a higher temperature.

  If hydrogen is used in rocket engines then oxygen is also required so that the hydrogen will burn.

  Steel making is another industry that needs oxygen.
- **Chlorine** is required for the manufacture of PVC, (polyvinyl chloride). It is used to sterilise water.

  Chlorine is used to make sodium chlorate(I) (hypochlorite), which is found in liquid domestic bleaches. It is also used in industry to bleach cotton, linen and wood pulp.
- **Carbon** is used as a fuel and to make the anodes used in the extraction of aluminium.

  In the form of charcoal, carbon is used for decolourising and deodorising. The open structure of the charcoal traps the molecules that give the colour or cause the smell.
- **Helium** is used to fill balloons. It is safer than hydrogen because it does not burn. Breathing mixtures for deep-sea divers can use helium in place of nitrogen to prevent the divers suffering from the 'bends'.
- **Neon** is used in fluorescent lighting to give a red-orange glow.

Uses of non-metal compounds

- **Carbon dioxide** is used in fire extinguishers for putting out electrical fires or burning liquids. It is dissolved under pressure to provide the fizz in soft drinks. It is used in its solid form as dry ice. The 'ice' is dry because it sublimes rather than melts and so does not leave any wetness.
- **Ammonia** is used directly or indirectly, via nitric acid, to make fertilisers. It is also required in the manufacture of some explosives and of nylon.
- **Sulphuric acid** is needed as the electrolyte in car batteries. It is also in great demand in industry, for example, in the manufacture of detergents, fibres and pigments.
- **Sulphur dioxide** is used as a bleach, for example for paper. It can be used as a preservative for dried fruit and as a fungicide where it kills unwanted microbes.
- **Nitric acid** is used mainly to make nitrate fertilisers, but a small percentage is needed in the manufacture of explosives.

Websites

www.gcsechemistry.com/ukop/htm

www.s-cool.co.uk

Exam questions

1 a) Copy and complete the table below which relates properties of non-metals to their uses.

Non-metal	Property	Use
hydrogen	produces energy on burning	
carbon		electrodes
chlorine	destroys harmful bacteria	

(3 marks)

b) The table below gives some information about four elements.

Element	Boiling point °C	Electrical conductivity in solid	in liquid	Behaviour in water
A	−269	none	none	does not dissolve
B	59	none	none	dissolves slightly
C	890	good	good	reacts rapidly
D	2567	good	good	does not dissolve or react

(i) How can you tell from the information in the table that A and B are non-metals?

(1 mark)

(ii) Which element A, B, C or D could be sodium?

Give a reason for your answer.

(2 marks)

c) Hydrogen, carbon dioxide and oxygen are gases that are easily prepared in the laboratory.

(i) Which of these gases turns limewater milky?

(1 mark)

(ii) Describe how you would test for oxygen gas.

(2 marks)

(iii) Using oxygen as an example explain the meaning of the term **diatomic gas**.

(2 marks)

d) When carbon dioxide is dissolved in water, an acidic solution is formed.

(i) Choose the pH value you would expect for this solution form this list.

1   5   7   9   13   *(1 mark)*

(ii) What acid is formed when carbon dioxide is dissolved in water?

(1 mark)

(iii) Why are scientists concerned about the increasing levels of carbon dioxide in the atmosphere?

(2 marks)

2 a) The first airships were filled with hydrogen gas. Now helium is used in airships.

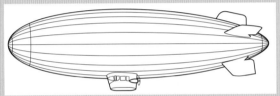

(i) What property do hydrogen and helium have in common which makes them useful in airships?

(1 mark)

(ii) Explain why helium is now used in airships and hydrogen is no longer used.

(2 marks)

(iii) Write a balanced symbol equation to show what happens when hydrogen burns in oxygen gas.

(2 marks)

b) Copper(II) oxide can react with hydrogen gas.

(i) Describe how you would carry out this reaction in the laboratory. You may use words and/or a diagram in your answer.

(2 marks)

(ii) What would you observe during this reaction of copper(II) oxide with hydrogen?

(2 marks)

(iii) Why is hydrogen gas described as a reducing agent when it reacts with copper(II) oxide?

(1 mark)

3 Water is essential for all life processes and is made up of **two** elements, hydrogen and oxygen. These two separate elements may be prepared in the laboratory using the apparatus below.

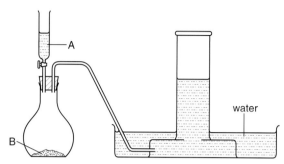

a) For the **safe** preparation of each gas identify solution A and solid B. Copy and complete this table.

	Solution A	Solid B
Hydrogen		
Oxygen		

(4 marks)

b) For each preparation give a **word** equation for the reaction taking place.

(2 marks)

c) Both hydrogen and oxygen are diatomic. Explain briefly what the term 'diatomic means'.

(2 marks)

d) Describe the tests used to identify the gases hydrogen and oxygen in the laboratory, including any observations made.

(4 marks)

e)

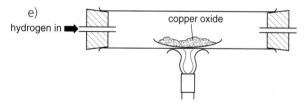

Hydrogen reacts with copper oxide in the apparatus shown above.

(i) Describe what would be **observed** in the reaction.

(3 marks)

(ii) Give a **symbol** equation for the reaction between hydrogen and copper oxide.

(2 marks)

(iii) Explain fully why the reaction between hydrogen and copper oxide is described as a **redox** reaction.

(5 marks)

f) Hydrogen is considered to be a clean fuel. Explain why hydrogen is a **clean fuel**.

(4 marks)

g) Give **one** use of hydrogen (other than as a fuel) and **one** use of oxygen.

(2 marks)

4 a) Sulphur is a yellow, non-metallic element which can exist as allotropes. Define what is meant by the term allotrope and name the allotropes of sulphur.

(5 marks)

b) The most important chemical that can be made from sulphur is sulphuric acid. Describe how sulphur is converted into sulphuric acid stating relevant conditions of temperature, pressure and catalyst as necessary. Write balanced, symbol equations where appropriate.

(12 marks)

c) Concentrated sulphuric acid shows reactions that are different from those of dilute sulphuric acid. Describe what you would observe when concentrated sulphuric acid is added to

(i) sugar *(3 marks)*

(ii) sodium chloride *(2 marks)*

Chapter 14

Energetics

By the end of this chapter you will:

➤ Know that materials can be decomposed by heating

➤ Understand the thermal decomposition of; hydrated copper(II) sulphate and metal carbonates including limestone and copper carbonate

➤ Understand the thermal cracking of hydrocarbons

➤ Be able to give examples of exothermic and endothermic reactions, for example, combustion, photosynthesis, dissolving, displacement, hydration of copper sulphate, neutralisation and electrolysis

➤ Recognise that energy transferred in a chemical reaction is associated with making and breaking bonds

Exothermic reactions

Fuels, such as coal or natural gas store chemical energy and when they burn they convert their chemical energy into heat energy.

$$CH_4 + 2O_2 \rightarrow CO_2 + 2H_2O + \textbf{heat energy}$$

In power stations this heat energy turns water into steam, which is used to turn turbines. The turbines turn generators to produce electricity.

A chemical reaction that gives out heat energy is called an **exothermic reaction**. The following are examples of exothermic reactions:

Neutralisation

hydrochloric acid + sodium hydroxide → sodium chloride + water
$HCl(aq)$ + $NaOH(aq)$ → $NaCl(aq)$ + $H_2O(l)$

Combustion

carbon + oxygen → carbon dioxide
$C(s)$ + $O_2(g)$ → $CO_2(g)$

Displacement reactions

zinc   + copper(II) sulphate →   zinc sulphate   + copper
$Zn(s)$ + $CuSO_4(aq)$ → $ZnSO_4(aq)$ + $Cu(s)$

magnesium +   sulphuric acid   → magnesium sulphate + hydrogen
$Mg(s)$ + $H_2SO_4(aq)$ → $MgSO_4(aq)$ + $H_2(g)$

Dissolving

Some substances dissolve exothermically, for example, sodium hydroxide:

$$NaOH(s) + H_2O(l) \rightarrow Na^+(aq) + OH^-(aq)$$

Hydration

Hydration of copper(II) sulphate.

Copper(II) sulphate reacts with water to form a hydrate and heat is given out:

copper(II) sulphate + water → hydrated copper(II) sulphate
$$CuSO_4(s) + 5H_2O(l) \rightarrow CuSO_4.5H_2O(s)$$

Endothermic reactions

While many chemical reactions are exothermic, there are also those which take in heat energy and these are called **endothermic reactions**.

Some examples of endothermic reactions are listed below.

Thermal decomposition

Thermal decomposition is the breaking down of a compound by heat. When limestone is heated strongly it breaks down into calcium oxide and carbon dioxide. Calcium oxide, commonly known as quicklime is a basic oxide and is an important chemical in agriculture. Farmers use it to neutralise acidic soil.

$$CaCO_3(s) \rightarrow CaO(s) + CO_2(g)$$

Copper(II) carbonate is a green solid that is readily broken down by heat to form black copper(II) oxide and carbon dioxide

$$CuCO_3(s) \rightarrow CuO(s) + CO_2(s)$$

Blue hydrated copper(II) sulphate also undergoes thermal decomposition forming white anhydrous copper(II) sulphate.

Hydrated copper(II) sulphate → anhydrous copper(II) sulphate + water
$$CuSO_4.5H_2O(s) \rightarrow CuSO_4(s) + 5H_2O(l)$$
blue solid   white solid

When water is added to white anhydrous copper(II) sulphate it turns blue because the reaction above is reversible . This reaction is used as a chemical test for water.

Cracking

This reaction is used to break large hydrocarbon molecules into smaller molecules. It is an important reaction in the oil industry and chemists use it to convert heavier hydrocarbons into more useful smaller hydrocarbon molecules (see pages 158–159). Thermal cracking takes place at temperatures around 1000°C while catalytic cracking takes place at around 500°C using a zeolite catalyst. Dodecane can be cracked to form ethene and the ethene is then used to make polythene.

dodecane → decane + ethene
$$C_{12}H_{26} \rightarrow C_{10}H_{22} + C_2H_4$$

Photosynthesis

Photosynthesis takes place in the leaves of green plants. Here chlorophyll acts as a catalyst to convert carbon dioxide and water in the presence of sunlight into sugars.

$$\text{energy (sunlight)} + \text{carbon dioxide} + \text{water} \rightarrow \text{glucose} + \text{oxygen}$$
$$\text{light} + 6CO_2(g) + 6H_2O(l) \rightarrow C_6H_{12}O_6(aq) + 6O_2(g)$$

Dissolving

When potassium nitrate crystals are dissolved in water, heat energy is taken in from the surroundings and the temperature of the water drops.

$$KNO_3(s) + H_2O(l) \rightarrow K^+(aq) + NO_3^-(aq)$$

Ammonium chloride also dissolves in water with a decrease in temperature.

Electrolysis

Electrolysis is another endothermic process which uses electricity to break down electrolytes (see page 97).

$$\text{lead bromide} \rightarrow \text{lead} + \text{bromine}$$
$$PbBr_2 \rightarrow Pb + Br_2$$

Questions

1   Explain the meaning of the following terms:

   a) exothermic   b) endothermic
   c) thermal decomposition   d) neutralisation
   e) cracking   f) electrolysis
   g) photosynthesis.

2   Complete the following equations for thermal decomposition;
   (i) $CuCO_3 \rightarrow CO_2 +$
   (ii) $CaCO_3 \rightarrow CaO +$
   (iii) $CuSO_4.5H_2O \rightarrow \quad +$

3   Complete the equations and state if the reactions are endothermic or exothermic

Reaction	Exothermic or endothermic
$KOH(aq) + HCl(aq) \rightarrow$	
$Zn(s) + H_2SO_4(aq) \rightarrow$	
$Light + 6CO_2(g) + 6H_2O(l) \rightarrow$	
$Fe(s) + CuSO_4(aq) \rightarrow$	

Energy level diagrams

So far several chemical reactions have been classified as endothermic or exothermic reactions depending on whether heat energy is taken in or given out.

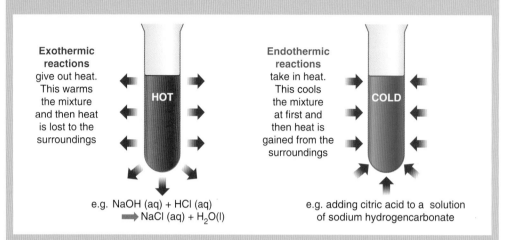

Figure 1 Exothermic and endothermic reactions

To help understand the energy changes that take place in a chemical reaction it is possible to draw energy level diagrams. Energy level diagrams are drawn for the combustion of carbon and the reaction of carbon with steam (Figure 2).

$$C(s) + O_2(g) \rightarrow CO_2(g) \qquad\qquad C(s) + H_2O(g) \rightarrow CO(g) + H_2(g)$$

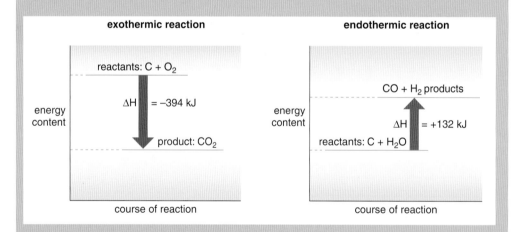

Figure 2 Energy level diagrams for exothermic and endothermic reactions

The energy level diagrams show the energy stored in the reactants relative to the energy stored in the products. The difference in energy between the reactants and products is given the symbol ΔH (delta H) and is the energy change for the reaction.

For the combustion of carbon we see that the reactants are losing energy to the surroundings. This is an example of an exothermic reaction because energy is **lost** to the surroundings ΔH is given a negative sign. The energy change is written as $\Delta H = -394$ kJ. This explains why the temperature of the surroundings increases.

For the reaction between carbon and steam it is seen that the products have a higher energy content that the reactants. In this case energy has been gained from the surroundings and ΔH is given a positive sign, $\Delta H = +132$ kJ. This explains why the temperature of the surroundings decreases.

Explaining energy changes in chemical reactions

Consider the reaction between hydrogen and chlorine to give hydrogen chloride gas as shown below:

$$H_2 \quad + \quad Cl_2 \quad \rightarrow \quad 2HCl$$
$$\text{H—H} \quad \text{Cl—Cl} \quad \text{H—Cl}$$
$$\text{H—Cl}$$

In the above reaction it is seen that the covalent bonds between the hydrogen atoms in the H_2 molecules and those between the chlorine atoms in the Cl_2 molecules must break and then new bonds must form between the hydrogen atoms and the chlorine atoms to give HCl molecules.

To break the covalent bonds in chlorine or hydrogen molecules requires energy and when new bonds are formed in the hydrogen chloride molecules, energy is given out. Bond breaking is an endothermic process as energy must be put in to break the bonds, while bond formation is an exothermic process as energy is given out when new bonds are formed.

Using bond energies it is possible to calculate whether the reaction between hydrogen and chlorine is exothermic or endothermic. From Table 1 it is seen that the bond energy for hydrogen is 436 kJ per mole, 242 kJ per mole for chlorine and 431 kJ per mole for hydrogen chloride.

Table 1 Bond energies for different bonds

Bond	Bond energy/kJ per mole
H—H	436
Cl—Cl	242
H—Cl	431
C—H	413
C—C	347
O=O	498
C=O	805
H—O	464

These bond energy values tell us how much energy is required to break or form one mole of bonds, for example, to break a mole of hydrogen bonds requires 436 kJ or when one mole of hydrogen bonds forms 436 kJ of energy is given out. Figure 3 shows how to calculate how much energy is required to break bonds in the reactant molecules and how much energy is formed when new bonds form

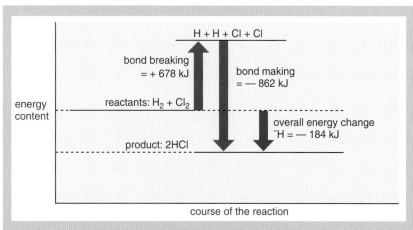

Figure 3 Calculating the energy change for bond breaking and making when one mole of hydrogen reacts with one mole of chlorine

We can now calculate the energy change for the reaction, ΔH

Energy required for breaking bonds in H_2 and Cl_2 = 678 kJ (436 + 242 kJ)
Energy given out when hydrogen chloride bonds form = 862 kJ (2×431 kJ)
Amount of energy given out = 862 − 678 kJ
= 184 kJ

as it is an exothermic reaction then ΔH = −184 kJ

The following energy level diagram can be used to represent this exothermic reaction.

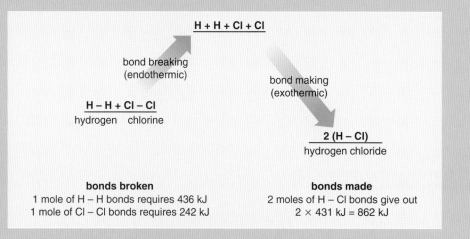

Figure 4 Energy level diagram for the reaction between hydrogen and chlorine

In the above diagram we can see that the reaction between hydrogen and chlorine is exothermic because there is more energy given out in the formation of the hydrogen chloride bonds than there is energy taken in, in the breaking of the bonds in hydrogen and chlorine.

From the study of energy level diagrams it is possible to make the following statements about energy changes in chemical reactions:

● If the energy required to break the bonds is greater than the energy given out when new bonds are formed then the reaction is endothermic and heat energy will be taken in from the surroundings.

● If the energy required to break the bonds is less than the energy given out when new bonds form then the reaction is exothermic and energy will be given out to the surroundings.

● If the energy required to break the bonds is equal to the energy given out when new bonds form then there is no overall change in energy and the temperature of the surroundings will remain constant.

Questions

4   The energy change for calcium carbonate breaking down to calcium oxide and carbon dioxide is $+1178$ kJ.

$$CaCO_3(s) \rightarrow CaO(s) + CO_2(g)$$

a) How can you tell from the energy change if the reaction is exothermic or endothermic?

b) From the value given for the energy change in the reaction, do the reactants have more or less energy than the products?

c) Draw an energy level diagram for the reaction, showing the relative positions of the reactants and products.

5   a) Use the bond energies in Table 1 to calculate the overall energy change in the reaction:

$$2H_2 + O_2 \rightarrow 2H_2O$$

b) Is the reaction exothermic or endothermic?

c) Draw an energy level diagram for the reaction.

(Remember H_2 is H—H and O_2 is O—O while water contains two O—H bonds.)

6   Use the website http://www.chem.umn.edu/outreach/EndoExo.html to obtain information that will help you design a pair of hand warmers. Give clear directions on how you will obtain energy from the chemicals you use and suggest how this energy can be released when required.

Websites

www.bbc.co.uk/schools/gcsebitesize/chemistry

www.gcsechemistry.com/ukop.htm

www.schoolscience.co.uk

Exam questions

1 Below are some examples of chemical processes.

Copy and complete the table to show which reactions are **exothermic**.

One has been done for you.

Reaction	Exothermic Yes/No
Turning water into steam	No
Photosynthesis	
Burning coal	
Neutralisation of acid and alkali	
Hydration of white copper sulphate to blue copper sulphate	

(4 marks)

2 Study the following symbol equations.

A $CaCO_3 \xrightarrow{\text{heat}} CaO + CO_2$

B $H_2O \xrightarrow{\text{heat}} H_2O$
(liquid)　　(gas)

C $CaCO_3 + H_2CO_3 \longrightarrow Ca(HCO_3)_2$

D $CuO + H_2 \longrightarrow Cu + H_2O$

a) Which equation, **A**, **B**, **C** or **D**, represents thermal decomposition?
(1 mark)

b) Which equation, **A**, **B**, **C** or **D**, does **not** represent a chemical reaction? Give a reason for your answer.
(2 marks)

3 This question is about copper sulphate.

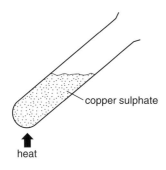

copper sulphate

heat

When blue copper sulphate crystals are heated an **endothermic** reaction takes place.

a) What do you understand by the term **endothermic** reactions?
(1 mark)

b) What colour do blue copper sulphate crystals turn when heated?
(1 mark)

c) How can you get the blue colour back again?
(1 mark)

4 a) Heat plays an important part in many chemical reactions. Some reactions require heat while others give out heat. What term is used to describe reactions which take in heat?
(1 mark)

b) The thermal decomposition of copper carbonate takes in heat.

(i) What is meant by the term 'thermal decomposition'?
(2 marks)

(ii) What **observations** would you make during this reaction?
(2 marks)

(iii) Write a **symbol** equation for the thermal decomposition of copper carbonate.
(2 marks)

c) The combustion of fuels is an important type of reaction that gives out heat. Write a **word** equation to show the **complete** combustion of the hydrocarbon propane.
(1 mark)

d) The combustion of fuels that have a high sulphur content gives rise to acid rain. Describe how acid rain is formed and name **two** ways in which our environment may be damaged by acid rain.
(5 marks)

5 a) What term is used to describe a reaction which gives out heat?

(1 mark)

b) When drops of water are added to anhydrous copper sulphate heat is given out.

(i) Describe what would be **observed** during this reaction.

(2 marks)

(ii) The addition of water to anhydrous copper sulphate is called

(1 mark)

c) Heat is also given out during the reaction of sodium hydroxide with hydrochloric acid.

(i) Give a balanced symbol equation for this reaction.

(2 marks)

(ii) What is the reaction of an acid with an alkali called?

(1 mark)

d) The redox reaction of zinc with copper sulphate solution also gives out heat.

(i) Describe what would be **observed** during this reaction.

(2 marks)

(ii) Give a balanced symbol equation for this reaction.

(2 marks)

6 Children like to eat a fizzy sweet substance called sherbet. When sherbet powder mixes with water an endothermic reaction takes place. The word equation for this reaction is as follows:

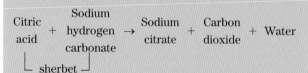

A pleasant **cooling** sensation is felt in the the mouth.

a) Explain in terms of the bonds how this cooling sensation happens.

(2 marks)

The graph below represents an **endothermic** reaction.

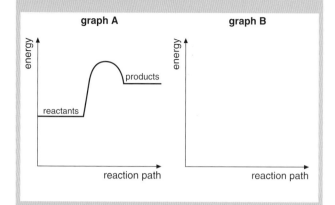

b) Draw and complete the similar graph, B, to represent an **exothermic** reaction.

(1 mark)

7 The diagram below shows the chemical changes which take place when methane burns in air.

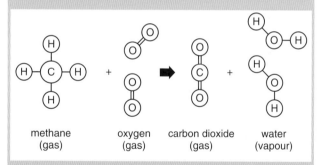

| methane (gas) | oxygen (gas) | carbon dioxide (gas) | water (vapour) |

Explain clearly in terms of bond-making and breaking how energy is produced during this reaction.

(3 marks)

Chapter 15

Organic Chemistry

Learning objectives

By the end of this chapter you will be able to do several things as listed under the relevant headings.

Sources of organic chemicals

➤ Recall that fossil fuels and biological substances are carbon-based compounds

➤ Recall that oil is the major source of organic chemicals and that the chemicals obtained from crude oil are hydrocarbons

➤ Understand the potential hazards of oil spillage

➤ Explain how chemicals are obtained from oil by fractional distillation and by cracking

➤ Understand the social, cultural, economic, environmental, health and safety and energy factors involved in organic chemistry processes

➤ Give examples of fossil fuels: natural gas, LPG, petrol, diesel, paraffin, candle wax, peat, lignite, coal and coke

➤ Recognise that fossil fuels are non-renewable resource

➤ Describe the combustion of fossil fuels (alkanes and alkenes) to produce carbon dioxide, water and heat energy

$$\text{Fossil fuel} + \text{oxygen} \rightarrow \text{carbon dioxide} + \text{water} + \text{heat energy}$$

➤ Recognise the problems caused by the combustion of hydrocarbons, limited to: the poisonous effects of carbon monoxide and the effects of carbon dioxide contributing to the 'greenhouse effect'

Homologous series

➤ Define a homologous series (for alkanes and alkenes) as one in which the chemicals have: the same general formula, similar chemical properties and a gradation in their physical properties

Alkanes

➤ Recall the names, molecular and structural formulae, and physical state of the first four alkanes

➤ Describe the complete combustion of alkanes and recognise their use as fuels

Alkenes

➤ Recall the names, molecular and structural formulae, physical state of the first two alkenes

➤ Understand that alkanes are saturated and alkenes are unsaturated

➤ Describe the combustion of alkenes, (manufacture of ethanol from ethene and steam, DAS), and the use of bromine water to distinguish between alkanes and alkenes

➤ Recognise the use of alkenes in the manufacture of commercially important addition polymers, polythene, PVC, (polypropene, DAS only) and (polystyrene, TAS only) and give uses of these polymers

➤ Recognise the problems associated with the disposal of the listed polymers by landfill and incineration

Ethanol

➤ Recall the name, molecular and structural formulae and physical state of ethanol

➤ Describe the formation of ethanol from: ethene and steam (reaction conditions are not required) and from fermentation (equation is not required)

➤ Describe the combustion of ethanol
➤ Recall the uses of ethanol as; a solvent, a fuel and in alcoholic beverages

➤ Understand the 'term safe drinking limit' and the possible harmful and beneficial effects of ethanol in alcoholic drinks

Ethanoic acid

➤ Recall the name, molecular and structural formulae of ethanoic acid
➤ Give examples of reactions of ethanoic acid as a typical dilute acid i.e. with metals, bases and carbonates

➤ Reaction with ethanol to form the ester, ethyl ethanoate (test-tube scale only)

➤ Recall the uses of ethanoic acid as a flavouring and a preservative

Ethyl ethanoate

➤ Recall the name, molecular and structural formulae of ethyl ethanoate

➤ Recall that ethyl ethanoate is used as a solvent and in flavourings

Fossil fuels

Most of the energy we use in our homes, industry, schools, offices and transport comes from **fossil fuels**. When fossil fuels burn, the chemical energy of the fuel is converted into heat energy. In the home the energy from fossil fuels is used mainly for heating, lighting and cooking while in industry much is used in the manufacture of materials, such as fertilisers, plastics, paints, metals, glass and other useful substances.

Methane gas, CH_4, which is the main component of natural gas, is an example of a fossil fuel and it burns producing heat energy as shown in the equation:

methane + oxygen → carbon dioxide + water + heat energy
$$CH_4(g) + 2O_2(g) \rightarrow CO_2(g) + 2H_2O(g) + \text{heat energy}$$

The general word equation for any fossil fuel burning is:

fossil fuel + oxygen → carbon dioxide + water + heat energy

Coal, oil, peat, lignite and natural gas are all examples of fossil fuels and were formed over many millions of years from dead animals and plants. Coke is also a fossil fuel that is made by heating coal in the absence of air.

Fossil fuels are **non-renewable** resources and this means that they are finite and once they are used up they cannot be replaced. It is estimated that the present supplies of crude oil will only last to around the end of this century. It is clear that in future there is a need for consumers to ensure that our supplies of crude oil are used more economically and realise the need to develop alternative energy sources.

Formation of fossil fuels

Oil and natural gas were formed from tiny animals and plants that lived in tropical seas while coal was formed from forests and vegetation growing in swamplands. Coal is a hard black solid which was formed in the Carboniferous period, about 300 million years ago, when trees and vegetation died and collected at the bottom of swamps. Flooding by seawater caused deposits of sand and mud to cover the decaying trees and vegetation in the swamps. Due to climatic changes and earth movements, layers of decaying wood and sediment built up, as shown in Figure 1. As a result of anaerobic decay (without oxygen), high pressure and temperature, the decaying matter was slowly compressed and converted into peat. Further pressure and heat compressed the peat to lignite and eventually to coal.

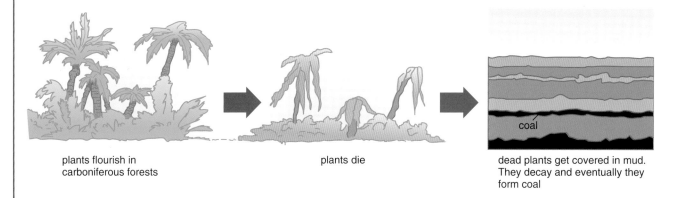

plants flourish in carboniferous forests

plants die

dead plants get covered in mud. They decay and eventually they form coal

Figure 1 The formation of coal

The hardness and percentage of carbon in a coal depends on the temperature and pressure at which it was formed. Anthracite is a hard coal which contains around 95% carbon and was formed at a higher pressure and temperature than bituminous coal, which contains around 70% carbon. The higher the percentage of carbon in a coal, then the greater the heat content of that coal.

Oil and natural gas were formed in a similar way to coal. Dead animals and plant life sank to the bottom of the seabed and were covered by mud, sand and silt. Under high pressure and temperature, anaerobic decay caused the dead animals and plants to be slowly converted into oil and natural gas. Figure 2 shows how oil and natural gas are obtained by drilling. It is seen that oil and natural gas are trapped in folds between layers of non-permeable rocks. When the long drill, from the drilling rig, bores through the top layer of impermeable rock into the oil and gas, the gas pressure forces oil through pipes to the surface.

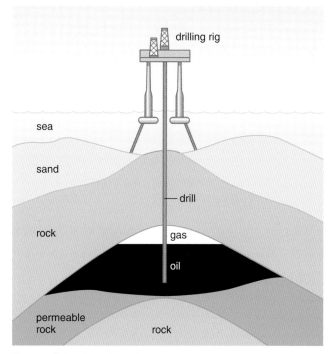

Figure 2 Drilling for oil and natural gas

Natural gas and crude oil are very important to our economy because they provide us with different fuels and a vast range of organic compounds for the chemical industry. These compounds can be used to make dyes, plastics, fertilisers, medicines, paints, lubricants and many other important classes of materials.

Although it is difficult to think how we would survive without oil, pollution of seas by oil can be a major problem. The transportation of crude oil in large tankers and the extraction of oil from the seabed have resulted in a number of serious oil spills and leakages. Generally oil spillages are cleaned up by spraying them with detergent. This breaks the oil up and allows bacteria to digest it.

One of the worst spills in recent years involved the Liberian-owned supertanker, *Sea Empress*, that ran aground at St Anne's Head, Wales in 1996. In that accident 70 000 tonnes of crude oil spilled out into the sea, in an area that was renowned for its bird sanctuaries and its National Marine Park. There was serious water pollution and birds, fish and mammals were severely affected. The oil on the birds hampered them from flying and caused a breakdown in their thermal insulation. A number of the birds were poisoned while others suffered severe irritation to the skin. Emergency teams were drafted in from Britain and Europe to help clean up the oil spill. Within weeks natural dispersal and the massive clean-up operation cleared most of the oil from the coastline.

Figure 3 Workers clean up the oil on the Pembrokeshire coast after the oil tanker Sea Empress ran aground in 1995

Questions

1 a) Name 5 fossil fuels
   b) what element do all fossil fuels contain?
   c) Complete the word equation to show what happens when a fossil fuel burns:

$$\text{Fossil fuel} + \text{oxygen} \longrightarrow$$

   d) Why are fossil fuels described as non-renewable resources?

2 Explain how the following fossil fuels were formed:

   a) crude oil and natural gas   b) coal

3 **IT:** Use a search engine to find out about major oil spillages at sea and use your findings to make a poster for display.

Crude oil

Crude oil is a viscous, dark brown liquid and is a mixture of many organic substances. Most of these substances are known as **hydrocarbons**. Some of the hydrocarbons have only a few carbon atoms in each molecule but there are some with over seventy carbon atoms in a molecule. A hydrocarbon is a substance which is composed of carbon and hydrogen only.

Fractional distillation

The hydrocarbons in crude oil have different boiling points and because of this they can be separated by using **fractional distillation**. During fractional distillation crude oil is separated into groups of hydrocarbons that have a similar number of carbon atoms in each molecule. Fractional distillation takes place in a steel tower as shown in Figure 4. The tower is very hot at the base and cooler at the top. Crude oil is fed in and heated in a furnace. The vapours are then fed into the fractionating tower. As the vapour mixture rises up the tower, different vapours condense at different levels and are separated. The small hydrocarbon molecules with the smallest number of carbon atoms in each molecule and with the lowest boiling points will rise to the top of the tower while the larger hydrocarbons with the highest boiling points will condense lower down in the tower.

Figure 4 shows that most of the fractions obtained from crude oil can be used as fuels. These fractions are LPG (liquid petroleum gas), petrol, naptha, paraffin, diesel, and candle wax. They are classified as fossil fuels as is crude oil. The properties of the different fractions can be investigated in the laboratory using the apparatus illustrated in Figure 5.

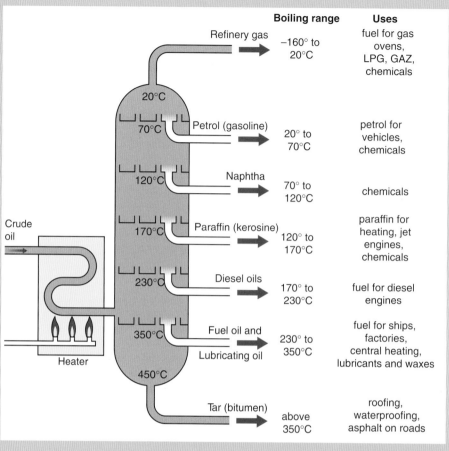

Figure 4 The products obtained by fractional distillation of crude oil and their major uses

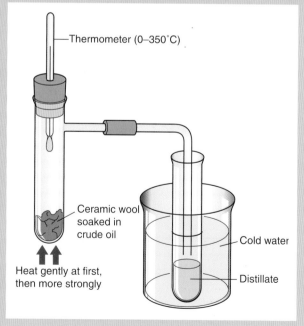

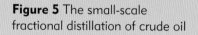

Figure 5 The small-scale fractional distillation of crude oil

Ceramic wool, soaked in crude oil is heated and the different fractions collect according to their boiling range. The fractions can then be tested for colour, viscosity and flammability. Typical results are given below:

Table 1 The properties of the fractions obtained from crude oil

Boiling range (°C)	Name of fraction	Colour	Viscosity	How does it burn?
20–70	petrol (gasoline)	pale yellow	runny	burns easily with a clean yellow flame
70–120	naphtha	yellow	fairly runny	burns quite easily with a yellow flame and some smoke
120–170	paraffin (kerosene)	dark yellow	fairly viscous	harder to burn, quite smoky flame when it does burn
170–270	diesel oils	brown	viscous	hard to burn, smoky flame when it does burn

The results show that as the number of carbon atoms in the fractions increases then:

● The fractions become more viscous or less runny.
● The colour of the fractions get darker.
● The boiling points of the fractions increase.
● The fraction become more difficult to burn and the flame becomes more smoky.

The properties of these fractions are taken into consideration when choosing a fuel for a particular purpose.

Cracking of hydrocarbons

In oil refineries it is found that there is a greater demand for the **lighter fractions**, refinery gas and petrol, than for some of the **heavier fractions**. This is due to the high demand for petrol, bottled gas and monomers such as propene and ethene to make plastics. Table 2 compares the percentages of the different fractions in crude oil with the percentage demand for them in everyday life.

Table 2 Comparison of the composition of crude oil with everyday demand for the fractions

Fraction	Percentage in crude oil	Percentage in everyday demand
Fuel gas	2	4
Petrol	6	22
Naphtha	10	5
Kerosine	13	8
Diesel oil	19	23
Fuel oil and bitumen	50	38

Fortunately chemists can convert the heavier fraction molecules into smaller molecules by **cracking** and this provides more of the lighter fractions to meet the demands of the petrochemicals industry. Figure 6 shows a catalytic cracking plant at an oil refinery.

Cracking is used to convert the heavier fractions of oil into more useful products by breaking large hydrocarbon molecules into smaller molecules. Cracking by means of high temperature (up to 1000°C) is called thermal cracking, and when a catalyst is used it is called catalytic cracking. A zeolite catalyst is generally used and this reduces the temperature to around 500°C. Cracking requires high temperatures because during the process strong covalent bonds between carbon atoms must be broken. To break these strong bonds a lot of energy is required.

Figure 6 The catalytic cracking plant at an oil refinery

Consider the cracking of decane ($C_{10}H_{22}$) in the naphtha fraction to produce petrol or octane (C_8H_{18}) and ethene (C_2H_4):

$$\text{decane} \rightarrow \text{octane} + \text{ethene}$$
$$C_{10}H_{22} \rightarrow C_8H_{18} + C_2H_4$$

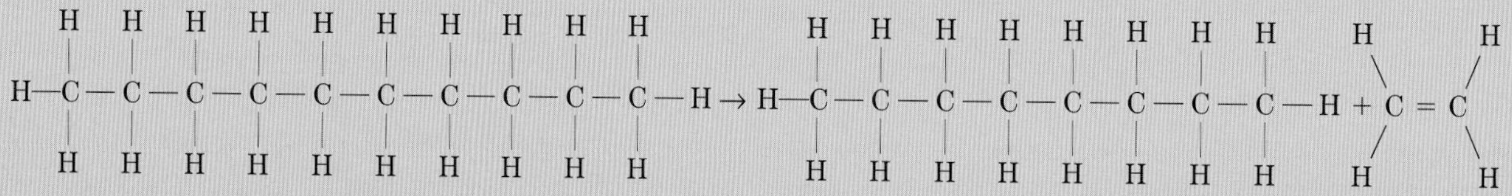

decane
(component of naphtha)

octane
(component of petrol)

ethene

Figure 7 Catalytic cracking of decane

Here decane in the naptha fraction produces two important smaller hydrocarbon molecules: petrol, which is used as a fuel for car engines, and ethene, which is used to make polythene in the plastics industry. Petrol is a mixture of alkanes that contain about eight carbon atoms. The petrol obtained from cracking is a better quality fuel than that obtained from fractional distillation. In the oil refinery, petrol obtained from cracking is blended with that obtained from fractional distillation to improve the petrol's quality.

During the cracking of decane (as shown in Figure 7) only one of the hydrocarbons produced is an alkane (octane) and the other is an alkene (ethene). When the molecular formula of octane, C_8H_{18} is taken away from decane, $C_{10}H_{22}$ the molecule remaining has a molecular formula C_2H_4 and is called ethene (see Alkenes, pages 1644–166). Because it contains a double bond between the two carbon atoms it is known as ethene not ethane (Figure 7).

Fossil fuels, combustion and air pollution

Air pollution is mainly caused by the burning of fossil fuels. During their combustion, harmful gases, such as carbon monoxide, carbon dioxide, sulphur dioxide and oxides of nitrogen are released into the atmosphere along with soot and smoke. Power stations and motor vehicles cause the greatest part of this pollution.

Carbon monoxide

Carbon monoxide is a colourless and odourless gas and is produced by the incomplete combustion of fossil fuels. This occurs when there is insufficient oxygen to form carbon dioxide and water. The principal source of carbon monoxide is from motor vehicles. Carbon monoxide is highly poisonous to animals, including humans. Haemoglobin is the red substance in blood and it carries oxygen around the body and releases it to the tissues. Carbon monoxide prevents haemoglobin transporting oxygen to the tissues because it displaces it from oxyhaemoglobin and forms carboxyhaemoglobin as shown in the equations below:

Haemoglobin + oxygen → oxyhaemoglobin
Oxyhaemoglobin + carbon monoxide → carboxyhaemoglobin + oxygen

This causes death because the tissues are starved of oxygen. As a result, it is important that in homes chimneys and flues are cleaned regularly and that gas and oil appliances are serviced as instructed by manufacturers. This will minimise the formation of carbon monoxide from the burning of fossil fuels and make the home a safer place to live in. Levels of carbon monoxide can be reduced in motor vehicles by using catalytic converters which change carbon monoxide to the less harmful carbon dioxide.

Greenhouse effect

Carbon dioxide levels are continually building up and this has contributed to global warming. It is called the **greenhouse effect**. We all know that the air inside a greenhouse is much warmer than that outside and this is caused by the glass trapping some of the sun's radiation as shown in Figure 8. In the same way the earth's atmosphere can trap some of the sun's radiation. By absorbing some of the sun's radiation the temperature of the earth's surface is maintained at around 15°C.

Carbon dioxide is very effective in the earth's atmosphere at keeping the earth warm. It is much more effective than oxygen or nitrogen, the main gases in the earth's atmosphere. Due to the excessive burning of fossil fuels the level of carbon dioxide in the atmosphere is continually rising and this is contributing to an increase in the greenhouse effect. This increased effect has led to the earth's temperature increasing and it is known as **global warming**.

● In the last 50 years the earth's temperature has increased by about 0.5°C. By the year 2100 it is predicted that there will be a temperature rise of 2°C. Although these appear to be small temperature increases such rises can have major long-term effects on our climate. Sea levels are rising due to water expansion and ice is melting at the Poles. This has caused flooding, especially in a number of lowland countries and coastal regions.

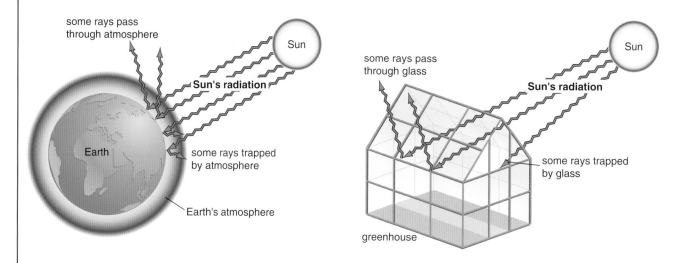

Figure 8 The greenhouse effect

● The greenhouse effect has also been used to explain decreased rainfall in African countries and this has led to crop failure.

● Higher temperatures than ever are being recorded in many areas. In Britain these higher temperatures have resulted in birds building their nests and laying their eggs earlier in spring.

Figure 9 Little egrets, attracted by the milder climate, are now breeding in the UK. Bluebells are now flowering earlier in the year

Questions

4 a) What is the meaning of the term **hydrocarbon**?

b) Explain how the different hydrocarbons in crude oil are separated by fractional distillation.

c) Petrol can be obtained from naptha by 'cracking'. Use an equation to explain the meaning of the term 'cracking' and state why cracking is such an important process in the petrochemicals industry.

5 The combustion of fossil fuels is a major cause of air pollution. Two of the main pollutants are carbon monoxide and carbon dioxide.

a) Explain why scientists are concerned about the increasing levels of carbon dioxide in the atmosphere.

b) Outline why carbon monoxide is a poison and describe ways of reducing the levels of this gas.

6 **IT:** Use the website http:www.schoolscience.co.uk and select the topic 'fossils into fuels' to find out more about crude petroleum. Read the sections on:

- sources of crude oil and natural gas
- processing and transporting crude oil
- looking to the future.

As you move through the various themes answer the questions highlighted. You can check how you have performed by clicking on show answers.

Alkanes

Alkanes are obtained from crude oil which is made up of many different hydrocarbons. Most of these hydrocarbons are alkanes. The simplest alkane is methane with a molecular formula of CH_4. In methane there are four strong, single covalent bonds that join the hydrogen atoms to the carbon atom. Table 3 gives the names, molecular and structural formulae, and physical state of the first four members of the alkanes.

Table 3 The first four members of the alkane

Name	Molecular formula	Structural formula	State at room temperature
Methane	CH_4	H | H—C—H | H	gas
Ethane	C_2H_6	H H | | H—C—C—H | | H H	gas
Propane	C_3H_8	H H H | | | H—C—C—C—H | | | H H H	gas
Butane	C_4H_{10}	H H H H | | | | H—C—C—C—C—H | | | | H H H H	gas

Figure 10 Propane is used in the burners of hot air balloons, and butane is used for camping gas stoves

Homologous series

All the members of the alkane series have the same general formula, C_nH_{2n+2} and it is seen that there is an increase in the boiling point as the number of carbon atoms in the molecule increases:

n	C_nH_{2n+2}	Boiling point °C	Melting point °C
1	CH_4	−162	−183
2	C_2H_6	−89	−173
3	C_3H_8	−42	−187
4	C_4H_{10}	−1	−135

Table 4

Compounds like the alkanes which have the same general formula, similar chemical properties and a gradation in physical properties are known as a **homologous series**.

Physical properties of the alkanes:

● The first four members are gases CH_4 to C_4H_{10}; from C_5H_{12} to $C_{17}H_{36}$ they are liquids, and from $C_{18}H_{38}$ onwards they are low melting point solids.

● They have low melting and boiling points as shown in the table above.

● Alkanes are insoluble in water: for example, petrol or oil do not dissolve in water.

● Alkanes are very unreactive because they only contain C—C and C—H single covalent bonds. These single bonds are very strong and difficult to break.

Chemical reactions of the alkanes

Combustion is the most important chemical reaction of the alkanes. In a plentiful supply of oxygen they burn to produce carbon dioxide, water and heat energy. Due to the large amount of heat given out when they burn, alkanes are good fuels.

● Natural gas is mainly methane gas

● Propane and butane are the main constituents of LPG (liquified petroleum gas). LPG is used in 'Calor gas' and 'Gaz' and these gases are commonly used in caravans, homes, boats and by campers.

The following equations show what happens when the earlier members of the alkanes burn to produce heat energy.

$$CH_4(g) + 2O_2(g) \rightarrow CO_2(g) + 2H_2O(l) + \text{heat energy}$$
$$2C_2H_6(g) + 7O_2(g) \rightarrow 4CO_2(g) + 6H_2O(l) + \text{heat energy}$$
$$C_3H_8(g) + 5O_2(g) \rightarrow 3CO_2(g) + 4H_2O(l) + \text{heat energy}$$
$$2C_4H_{10}(g) + 13O_2(g) \rightarrow 8CO_2(g) + 10H_2O(l) + \text{heat energy}$$

In a limited supply of oxygen, a mixture of carbon dioxide, water, carbon monoxide and soot forms. As carbon monoxide has no smell and is extremely toxic it is dangerous to burn alkanes where there is insufficient oxygen.

Questions

7 a) What is the meaning of the term 'homologous series'?

   b) Give the names, molecular formulae and the structural formulae for the first four members of the alkane series.

8 Natural gas is mainly methane while the main constituents of calor gas are butane and propane

   a) Write balanced symbol equations to show the combustion of methane and butane

   b) In terms of bond making and bond breaking explain why burning butane in oxygen is an exothermic reaction.

Alkenes

Alkenes are hydrocarbon molecules that contain a carbon–carbon double bond. The first two members of the alkenes are ethene, C_2H_4 and propene, C_3H_6. From the molecular formulae of ethene and propene it is seen that the general formula for alkenes can be written as C_nH_{2n}. Ethene and propene are important starting materials in the chemical industry because they are used to make useful polymeric materials such as polythene and polypropene.

Manufacture of alkenes

In the petrochemical industry ethene and propene are both manufactured by the cracking of alkanes obtained from crude oil. Earlier in Figure 6 we saw how ethene could be obtained by the cracking of decane:

$$\text{decane} \rightarrow \text{octane} + \text{ethene}$$
$$C_{10}H_{22} \rightarrow C_8H_{18} + C_2H_4$$

Physical properties

Alkenes, like alkanes, have low melting points and boiling points and are insoluble in water. Like ethane and propane, ethene and propene are both colourless gases.

Chemical reactions

Unlike alkanes, alkenes are much more reactive because they contain a carbon–carbon double bond in their structure. Ethene and propene are two of the most important chemicals in the petrochemicals industry because they react with many different substances to give a variety of useful chemicals.

$$\begin{array}{ccc}
H & & H \\
\diagdown & & \diagup \\
& C = C & \\
\diagup & & \diagdown \\
H & & H
\end{array}
\qquad
\begin{array}{ccc}
H & & CH_3 \\
\diagdown & & \diagup \\
& C = C & \\
\diagup & & \diagdown \\
H & & H
\end{array}$$

ethene   propene

Compounds like ethene and propene with a carbon–carbon double bond are said to be **unsaturated hydrocarbons**, while alkanes without a carbon–carbon double bond are said to be **saturated hydrocarbons**

The following reactions illustrate the greater reactivity of alkenes:

Combustion

Alkenes undergo combustion in the same way as alkanes.

ethene $+$ oxygen $\rightarrow$ carbon dioxide $+$ water $+$ heat energy
$C_2H_4(g) + 3O_2(g) \rightarrow 2CO_2(g) + 2H_2O(l) +$ heat energy
propene $+$ oxygen $\rightarrow$ carbon dioxide $+$ water $+$ heat energy
$2C_3H_6(g) + 9O_2(g) \rightarrow 6CO_2(g) + 6H_2O(l) +$ heat energy

Alkenes unlike alkanes, are not used as fuels because it is more economical to use them in the production of organic chemicals. Examples are ethanol, ethanoic acid, plastics, fibres, solvents, detergents and many other useful products of the petrochemicals industry.

Hydration of ethene

Ethene reacts with steam to produce ethanol. The reaction in which water is added onto the ethene molecule is called hydration. The reaction cannot be carried out in school laboratories because high pressure is needed. The conditions required for the reaction are: a phosphoric(v) acid (H_3PO_4) catalyst, a temperature of $330°C$ and a pressure of 70 atmospheres.

ethene $+$ steam $\rightarrow$ ethanol
$C_2H_4(g) + H_2O(g) \rightarrow C_2H_5OH(g)$

When water adds to the double bond of an alkene, it is called an **addition reaction**

Reaction of alkenes with bromine

Like water, bromine adds to the double bond of alkenes. If ethene or propene is bubbled into a solution of bromine water, the reddish/brown colour of the bromine goes colourless.

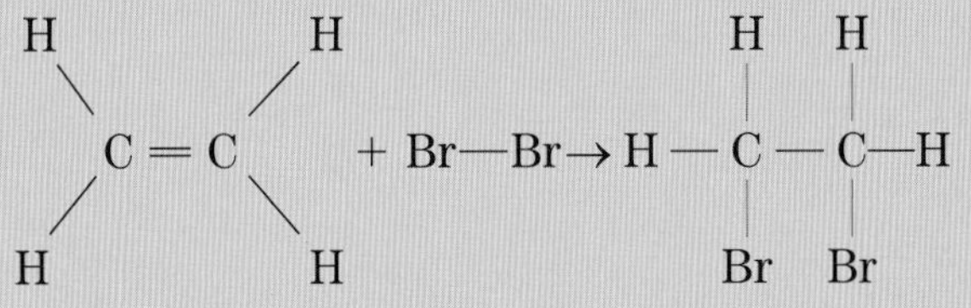

$$
\begin{array}{ccc}
\text{H} & & \text{H} \\
\diagdown & & \diagup \\
& \text{C} = \text{C} & \quad + \text{Br—Br} \rightarrow \\
\diagup & & \diagdown \\
\text{H} & & \text{H}
\end{array}
\qquad
\begin{array}{c}
\text{H} \quad \text{H} \\
| \quad\;\; | \\
\text{H} - \text{C} - \text{C—H} \\
| \quad\;\; | \\
\text{Br} \;\; \text{Br}
\end{array}
$$

ethene   +   bromine → 1,2-dibromoethane
C_2H_4 + Br_2 $C_2H_4Br_2$

The numbers 1 and 2 show that a bromine atom has added to each of the carbon atoms in the ethene molecule. Bromine water is used to distinguish between alkanes and alkenes. Alkenes decolourise bromine water while alkanes do not.

Polymerisation of alkenes

In chapter 4 we looked at the properties and uses of a number of **polymers**. Here we will study the formation of polymers from alkenes. The word 'polymer' comes from Greek and means many parts. Chemists use the word polythene to describe a molecule that has been made from many ethene molecules chemically joined or added together in long chain molecules. The reactive small molecules which make up the polymer are called **monomers**. For alkenes, the polymerisation reaction is also called addition polymerisation.

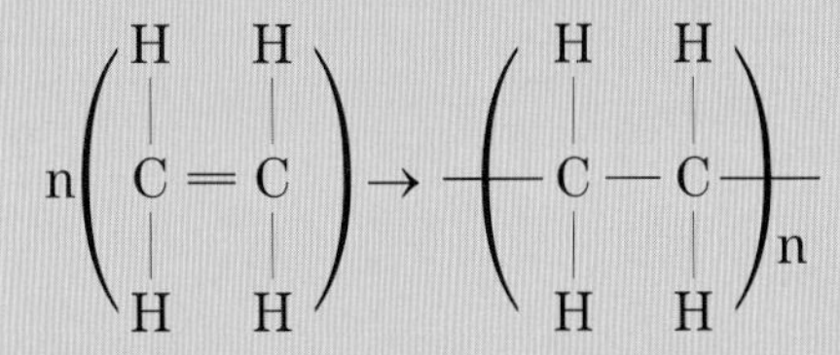

$$
n\left(\begin{array}{c} \text{H} \quad \text{H} \\ | \quad\;\; | \\ \text{C} = \text{C} \\ | \quad\;\; | \\ \text{H} \quad \text{H} \end{array}\right) \rightarrow \left(\begin{array}{c} \text{H} \quad \text{H} \\ | \quad\;\; | \\ \text{C} - \text{C} \\ | \quad\;\; | \\ \text{H} \quad \text{H} \end{array}\right)_n
$$

Figure 11 General equation for polymerisation of ethene, where n has a value between 500 and 1500

Polythene is made by heating ethene at high pressure in the presence of a catalyst.

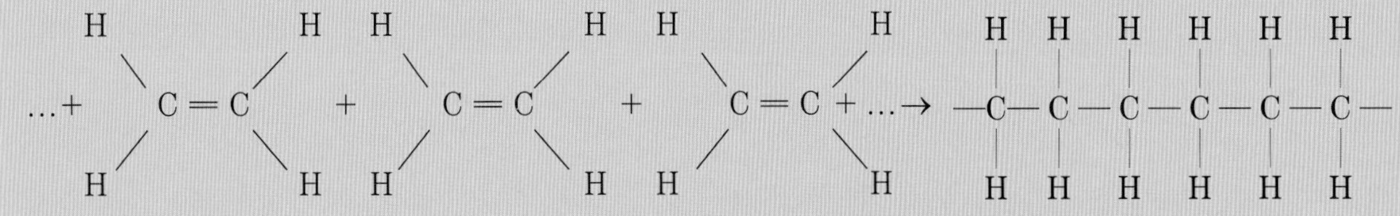

Figure 12 How ethene monomers join up to form the long chain of polythene

More about polymers

While polythene is a very useful addition polymer, three other common and important polyalkenes are; polyvinyl chloride (PVC), polypropene and polystyrene. The equations for formation of these polymers from their monomers are:

$$n\left(\begin{array}{cc} H & Cl \\ | & | \\ C = C \\ | & | \\ H & H \end{array}\right) \rightarrow \left(\begin{array}{cc} H & H \\ | & | \\ -C - C- \\ | & | \\ H & H \end{array}\right)_n \text{PVC}$$

$$n\left(\begin{array}{cc} H & CH_3 \\ | & | \\ C = C \\ | & | \\ H & H \end{array}\right) \rightarrow \left(\begin{array}{cc} H & CH_3 \\ | & | \\ -C - C- \\ | & | \\ H & H \end{array}\right)_n \text{Polypropene}$$

$$n\left(\begin{array}{cc} H & C_6H_5 \\ | & | \\ C = C \\ | & | \\ H & H \end{array}\right) \rightarrow \left(\begin{array}{cc} H & C_6H_5 \\ | & | \\ -C - C- \\ | & | \\ H & H \end{array}\right)_n \text{Polystyrene}$$

Some of the uses of the polyalkenes are given in Table 5 and Figure 13.

Figure 13 Some uses of polymers

Table 5 Properties and uses of some addition polymers

Monomer	Addition polymer	Properties	Uses
Polythene	$\left(\begin{array}{cc} H & H \\ \|&\| \\ -C-C- \\ \|&\| \\ H & H \end{array}\right)_n$	Light, flexible and resistant to attack by acids and alkalis	Cling film, plastic bags, bottles, buckets and basins
Polypropene	$\left(\begin{array}{cc} H & CH_3 \\ \|&\| \\ -C-C- \\ \|&\| \\ H & H \end{array}\right)_n$	Light, flexible, durable and resistant to attack by acids and alkalis	Plastic crates, ropes, carpets and packaging material
PVC	$\left(\begin{array}{cc} H & Cl \\ \|&\| \\ -C-C- \\ \|&\| \\ H & H \end{array}\right)_n$	Tough, durable, water proof and good insulator	Electric cables, guttering, drain pipes and umbrellas
Polystyrene	$\left(\begin{array}{cc} H & C_6H_5 \\ \|&\| \\ -C-C- \\ \|&\| \\ H & H \end{array}\right)_n$	Low density and a good insulator	Insulation, ceiling tiles, packaging and disposable cups

While the manufacture of polymers provides us with many important man-made materials there are serious problems associated with their disposal. It is not unusual to see plastic litter in the countryside, towns and even on our scenic beaches. Unlike natural polymeric materials such as paper, cotton and wool, most synthetic plastics are non-biodegradable and micro-organisms cannot break them down.

The fact that it is relatively cheap to produce plastics from crude oil further adds to the litter problem. As plastics are cheap most people tend to dump them, rather than recycle them. Recycling plastics is costly as it is expensive to have the waste collected and sorted. The recycling process also involves high energy costs.

The three main methods of dealing with waste are dumping in a landfill site, burning in an incinerator and recycling (see Figure 14).

Figure 14 a) Plastic waste has traditionally been dumped in landfill, b) it can now also be burned to produce electricity or c) recycled, such as these pellets of polythene

In recent years most of the old quarry sites available for landfill dumping have been used up and dealing with dumping plastics by this method is now relatively expensive, especially with the recent introduction of government landfill tax on waste.

Incineration schemes burn waste plastics and the heat generated can be made to serve a useful purpose, either directly or to generate power. The main problems associated with incineration are pollution from the products of burning and the destruction of many resources that could have been recycled. During the incineration process poisonous gases such as hydrogen chloride, carbon monoxide and hydrogen cyanide are given off and it is of extreme importance that these compounds are controlled and removed before any waste gases are released into the atmosphere.

Recycling of plastic waste is now becoming more and more popular. Figure 15 shows how thermosoftening and thermosetting plastics are recycled. Recycling thermosoftening plastics is carried out by melting the waste plastic and then remoulding it. Large quantities of the recycled plastic are used to make sheeting and plastic bags. Recycling of thermosetting plastic is much more difficult than the recycling of thermosoftening polymers because they do not melt on heating but burn or char.

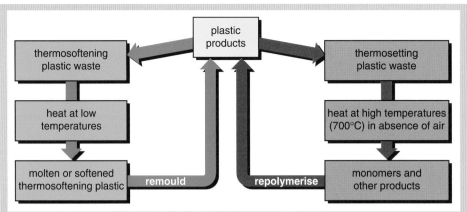

Figure 15 Recycling plastics

To recycle a thermosetting plastic, the plastic is first pyrolysed (heated in the absence of air) at around 700°C. Under these conditions the polymer breaks down forming the monomers it was originally made from. The monomers are collected and are then polymerised to give a new thermosetting polymer.

More recently chemists have developed new **biodegradable plastics** for film, packaging and containers and in future years this could be an important method to help resolve the present environmental difficulty of the disposal of plastics.

Figure 16 A biodegradable plastic bag made from maize starch

Questions

9   a) Give the names, molecular formulae and structural formulae of the first two members of the alkene series.
    b) Why is there no member in the alkene series with just one carbon atom in the molecule?
    c) Explain why alkene molecules are described as unsaturated molecules while alkanes are described as saturated.
    d) Write a balanced symbol equation to show how propene reacts with oxygen.
    e) Describe a chemical test you could carry out to show that an organic chemical was ethene and not ethane.

10 Alkenes are important chemicals that can be used to make alcohols and thermosoftening polymers

    a) Describe how ethene is converted into ethanol. Give a symbol equation with state symbols.
    b) What type of reaction is represented by: Ethene → polythene?
    c) Draw the structure of polythene and give two uses of the polymer.
    d) Polythene is a thermosoftening polymer. What is the meaning of the term thermosoftening polymer?
    e) Give one use of a thermosetting plastic
    f) Why is polystyrene a suitable material to use in the manufacture of ceiling tiles?
    g) List the advantages and disadvantages of recycling plastics.

Ethanol, C₂H₅OH

Ethanol is the second member of the homologous series known as the **alcohols**. It is a colourless liquid which boils at 79°C and has the following molecular structure:

$$H-\overset{\overset{\displaystyle H}{|}}{\underset{\underset{\displaystyle H}{|}}{C}} - \overset{\overset{\displaystyle OH}{|}}{\underset{\underset{\displaystyle H}{|}}{C}} - H$$

Its molecular formula is sometimes written as CH_3CH_2OH. It has a molecular structure similar to that of ethane except that one of the hydrogen atoms has been replaced with an –OH group. Ethanol is the alcohol that is contained in whisky, wine and other alcoholic drinks.

Making ethanol

Ethanol can be made either by the hydration of ethene which is abundantly available from cracking in the petrochemicals industry or by fermentation of sugar.

Hydration of ethene

$$\underset{C_2H_4(g)}{ethene} + \underset{H_2O(g)}{steam} \rightarrow \underset{C_2H_5OH(g)}{ethanol}$$

The conditions required for the reaction are; a phosphoric(v) acid (H_3PO_4) catalyst, a temperature of 330°C and a pressure of 70 atmospheres.

Fermentation

For hundreds of years ethanol has been made from the two carbohydrates, sugar and starch. If yeast is added to sugar or starch solution and the mixture left in a warm place, enzymes in the yeast convert the carbohydrates to alcohol and carbon dioxide. This breakdown of carbohydrates to form ethanol and carbon dioxide is known as fermentation. The following equations show how a sugar is broken down to glucose and then to ethanol during fermentation:

$$\underset{sugar}{C_{12}H_{22}O_{11}} + H_2O \rightarrow \underset{glucose}{2C_6H_{12}O_6}$$

$$\underset{glucose}{C_6H_{12}O_6} \rightarrow \underset{ethanol}{2C_2H_5OH} + 2CO_2$$

Figure 17 shows how fermentation can be carried out in a small scale in the laboratory.

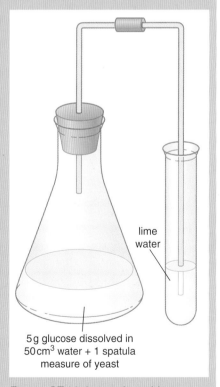

lime water

5 g glucose dissolved in 50 cm³ water + 1 spatula measure of yeast

Figure 17 Making ethanol by fermentation

Alcoholic drinks are made by fermentation. The fermentation of barley produces a weak solution of ethanol, (about 3% in ethanol), and this is how beer is made, while wine is made by the fermentation of grapes and produces a solution that contains about 13% ethanol. Spirits such as rum, brandy whisky and gin have a higher ethanol concentration and these are obtained by fractional distillation.

Many people socialise by going out with their friends for a drink. They find it enjoyable and relaxing. Nevertheless, it has to be realised that drinking in excess is harmful to our health causing serious illnesses such as heart disease, cirrhosis of the liver, damage to the nervous system, brain damage and alcoholism. Alcohol, like many other drugs is addictive and those suffering from alcoholism find it difficult to live normal lives. Very often it leads to family and work problems. Experts agree that to reduce diseases and other problems caused by alcohol people should stick to a **safe** drinking limit. This safe limit has been set at four units of alcohol per day for men and two units for women.

Small amounts of alcohol can affect people's judgment and their ability to drive, so it is important that people do not drive after consuming alcohol. Around 30 percent of accidents on the roads are caused by alcohol, and many of these by young drivers. The legal limit for driving a car is 80 milligrams of alcohol per hundred cubic centimetres of blood (80 mg/100 cm^3 blood).

Properties of ethanol: combustion

Ethanol burns with a pale yellow flame forming carbon dioxide and water and giving out heat energy:

$$C_2H_5OH(l) + 3O_2(g) \rightarrow 2CO_2(g) + 3H_2O(g) + \text{heat energy}$$

Like alkanes, ethanol is an important fuel. Countries such as Brazil that depend on imported oil have started producing ethanol as a fuel. They produce the alcohol by the fermentation of cane sugar and then use it in motor vehicle engines. Up to 10 percent ethanol is added to the petrol for car engines.

The climate and the availability of land for growing cane sugar make Brazil an ideal country for ethanol production. Ethanol has also the advantage that it is a cleaner fuel than petrol and produces less pollution from carbon particles and carbon monoxide. It has no sulphur impurities and does not produce sulphur dioxide, one of the main gases responsible for acid rain. The more ethanol that countries like Brazil can produce then the less need they have to import oil from other countries.

Uses of ethanol

The use of ethanol as a fuel and in alcoholic drinks has been discussed. Ethanol is also an important solvent and is used in cosmetics, toiletries, lacquers, glues, paints and inks.

Ethanoic acid, an organic acid

Ethanoic acid is an organic acid and a member of the homologous series called carboxylic acids. It is a colourless liquid with a strong, sharp vinegary smell and sour taste. Vinegar, which is used as a flavouring and preservative in foods, is a solution of ethanoic acid in water.

Other organic acids that we are familiar with in everyday life are citric acid, which is present in citrus fruits such as oranges, ascorbic acid or vitamin C, which is found in fresh fruit and vegetables and aspirin which is a pain killer. The simplest carboxylic acid is formic acid or methanoic acid and this is the stinging liquid in nettles and ants' sting while butanoic acid is one of the carboxylic acids present in human sweat.

The molecular formula of ethanoic acid is CH_3CO_2H, sometimes written as CH_3COOH, while the structural formula is:

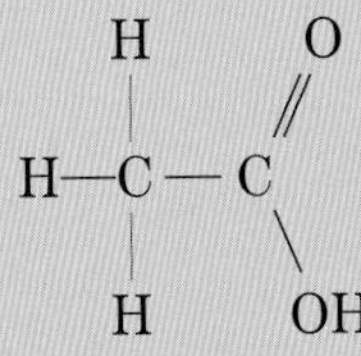

Figure 18a
Structural formula of ethanoic acid

Figure 18b All these foods contain vinegar (ethanoic acid)

Properties of ethanoic acid

Reaction as an acid

Ethanoic acid like other organic acids is a weak acid and reacts with metals, bases and carbonates:

with the more reactive metals it produces a salt and hydrogen

ethanoic acid + magnesium → magnesium ethanoate + hydrogen
$$2CH_3COOH + Mg \rightarrow (CH_3COO)_2Mg + H_2$$

with bases it produces a salt and water

ethanoic acid + sodium hydroxide → sodium ethanoate + water
$$CH_3COOH + NaOH \rightarrow CH_3COONa + H_2O$$

ethanoic acid + calcium oxide → calcium ethanoate + water
$$2CH_3COOH + CaO \rightarrow (CH_3COO)_2Ca + H_2O$$

with carbonates it produces a salt, carbon dioxide and water

ethanoic acid + potassium carbonate → potassium ethanoate + carbon dioxide + water

$$2CH_3COOH + K_2CO_3 \rightarrow 2CH_3COOK + CO_2 + H_2O$$

Did you know?

Human sweat contains carboxylic acids. Dogs can track people because they can detect differences in the carboxylic acids in our sweat

Esterification Reaction

Carboxylic acids react reversibly with alcohols to produce sweet smelling liquids called **esters**. The process is called **esterification**

ethanoic acid	+	ethanol		ethyl ethanoate	+	water
CH_3COOH	+	C_2H_5OH	$\rightleftharpoons$	$CH_3COOC_2H_5$	+	H_2O

The reaction can be carried out in a test tube by mixing ethanol and ethanoic acid and adding a few drops of concentrated sulphuric acid, which acts as a catalyst. If the mixture is gently heated, it is not long until the sweet smell of the ester, ethyl ethanoate is observed.

Esters are important commercial organic compounds because they are used as solvents and flavourings. Esters are commonly used as solvents for glues and the following table lists four examples of esters that are used in artificial flavourings:

Table 6 Esters used in flavourings

Ester	Artificial flavouring
Ethyl butanoate	Pineapple
Methyl butanoate	Apple
Propyl ethanoate	Pear
Pentyl ethanoate	Banana

Questions

11 Ethanol can be used as an alternative fuel to petrol.

   a) Give three reasons why ethanol is an important fuel.
   b) Give the molecular formula of ethanol and draw its structural formula.
   c) Describe two ways of making ethanol.
   d) Write a balanced symbol equation to show how ethanol burns in oxygen and describe the colour of the flame.
   e) give three uses of ethanol.

12 **IT:** Use a search engine to find out about ethanol as a fuel for the future.

13 Ethanoic acid is an organic acid and is a member of the homologous series known as the carboxylic acids.

   a) What do you understand by the term 'homologous series'?
   b) Name the organic acid found in citrus fruits like lemons.
   c) What is the common name for ethanoic acid?
   d) Write the molecular formula for ethanoic acid and draw its structural formula.
   e) Write word and balanced symbol equations to show how ethanoic acid reacts with:
      (i) magnesium   (ii) calcium oxide   (iii) potassium hydroxide
      (iv) sodium carbonate.
   f) Give two uses of ethanoic acid

14 Ethyl ethanoate is a sweet-smelling compound formed by the reaction of ethanoic acid with ethanol.

   a) Ethyl ethanoate is a member of which homologous series?
   b) Name the catalyst used in the above reaction to make ethyl ethanoate.
   c) Give the molecular formulae for ethanol, ethanoic acid and ethyl ethanoate.
   d) Write a balanced symbol equation to show how ethyl ethanoate is formed.
   e) Draw the structural formula of ethyl ethanoate.
   f) Give two uses of ethyl ethanoate.

Websites

www.bbc.co.uk/schools/revision

www.s-cool.co uk

Exam questions

1 The table below gives information about some of the hydrocarbon fuels obtained by fractional distillation of crude oil.

Fuel	Boiling range (°C)	Number of carbon atoms in each molecule
Petroleum gas	below 25	1–4
Petrol	40–100	4–12
Paraffin	150–240	9–16
Diesel	220–250	15–25

a) (i) What is a hydrocarbon?

(*2 marks*)

   (ii) The fuels listed above are all fossil fuels. Give **one** other example of a fossil fuel.

(*1 mark*)

   (iii) Explain how the different fuels are separated from each other by fractional distillation.

(*2 marks*)

   (iv) What is the name of the process used to break up long chain molecules in order to form more useful shorter chain molecules.

(*1 mark*)

b) Ethene and propene are hydrocarbons which belong to the same homologous series.

   (i) Name the homologous series to which ethene and propene belong.

(*1 mark*)

   (ii) Draw the structural formula of ethene and propene.

(*2 marks*)

c) Ethene burns in oxygen as shown in the equation below.

$$C_2H_4 + O_2 \rightarrow CO_2 + H_2O$$

   (i) Copy and balance the above equation

(*1 mark*)

   (ii) In terms of bond breaking and bond making explain why the burning of ethene in oxygen is an exothermic reaction.

(*3 marks*)

d) Four important organic chemicals are listed below.

        ethanol
        ethanoic acid
        ethyl ethanoate
        polypropene

   (i) Which **one** of these four chemicals is a solid at 25°C?

(*1 mark*)

   (ii) Which **one** of these four chemicals is an ester with a sweet smell?

(*1 mark*)

   (iii) Copy and complete the equation below which shows how ethanol can be manufactured from ethene. Include the missing state symbol.

$$C_2H_4(g) + \qquad (\ ) \rightarrow C_2H_5OH(g)$$

(*2 marks*)

   (iv) Describe any chemical test which you could use to show that an organic chemical was ethanoic acid and not ethanol.

(*3 marks*)

2 a) Ethene is a very useful chemical which can be burned as a fuel or used in the manufacture of polythene and ethanol.

              ↗ Polythene
ETHENE → Heat energy
           ↘ Ethanol → Ethanoic acid

   (i) Give **two** uses of polythene.

(*2 marks*)

   (ii) What type of reaction is represented by:

        ethene → polythene?

(*1 mark*)

(iii) Heat energy is produced when ethene burns in air.

Write a balanced symbol equation for the complete combustion of ethene.

(2 marks)

(iv) Ethene and propene are the first two members of a homologous series. What do you understand by the term '**homologous series**'?

(2 marks)

b) Ethanoic acid undergoes the typical reactions of an acid and also reacts with ethanol to form an ester.

(i) Draw the **structural** formulae of ethanol and ethanoic acid.

(2 marks)

(ii) Give **one** use of ethanol

(1 mark)

(iii) Describe the appearance of the ester, ethyl ethanoate.

(2 marks)

(iv) Name the products of the reaction of ethanoic acid with magnesium.

(2 marks)

3 a) Carbon is unique in its ability to form a very wide range of compounds in which carbon atoms are bonded to other carbon atoms.

(i) Draw structural formulae to show all the bonds in the compounds listed below.

ethane   propane
ethene   propene   *(8 marks)*

(ii) The compounds listed in a)(i) belong to two different homologous series. Copy and complete the table below naming the two series to which these compounds belong and placing each compound into the correct series.

Name of series	Compounds
1	
2	

(6 marks)

(iii) Describe a chemical test which would enable you to distinguish between ethane and ethene. Also state the results of the test.

(5 marks)

b) Ethene is a more useful compound than ethane. One use of ethene is to make the polymer polythene. Write an equation to show how ethene is converted into polythene.

(2 marks)

c) Another important polymer is PVC. Give **two** uses of this polymer and state one property of PVC which makes it suitable for each use.

(4 marks)

d) Ethene may also be converted into ethanol by direct combination with water using a catalyst. Write a balanced, symbol equation for this reaction.

(2 marks)

e) Ethanol reacts with ethanoic acid. State the conditions under which this reaction occurs, write a balanced, symbol equation for the reaction, name the compound formed and draw its full structural formula.

(8 marks)

4 a) Petrol is a fossil fuel, obtained from oil by fractional distillation.

(i) Name the element which all fossil fuels contain

(1 mark)

(ii) Why can petrol be described as a non-renewable energy source?

(1 mark)

(iii) Give two other examples of non-renewable energy sources.

(2 marks)

(iv) Explain how petrol is separated from the other fractions in crude oil by distillation.

(2 marks)

b) Copy and complete the table below which gives information about some alkanes and alkenes.

Name	Molecular formula	Structural formula
Methane	CH_4	H | H—C—H | H
Ethane	C_2H_4	
	C_3H_6	H H H \ | / H—C = C — C — H / \ H H
Butane		H H H H | | | | H—C — C — C — C —H | | | | H H H H

(*3 marks*)

c) Ethanol, C_2H_5OH, is an alternative fuel to petrol.

(i) Give one **other** use of ethanol.
(*1 mark*)

(ii) Draw the structural formula of ethanol.   (*1 mark*)

(iii) Write a symbol equation to show how ethanol can be manufactured using ethene and steam.
(*1 mark*)

(iv) The combustion of ethanol is an exothermic reaction. The equation of this reaction is shown below.

$C_2H_5OH + 3O_2 \rightarrow 3H_2O + 2CO_2$

Explain with reference to bond making and breaking why the combustion of ethanol is exothermic.   (*3 marks*)

5 a) Crude oil is a source of compounds called alkanes.

(i) Explain how fractional distillation separates the compounds found in crude oil.
(*2 marks*)

(ii) Name the first member of the alkanes and draw its structural formula.
(*2 marks*)

(iii) Alkanes can be used as fuels. Write a balanced chemical equation for the complete combustion of propane, C_3H_8.
(*2 marks*)

b) Alkenes can be made by cracking alkane molecules.

(i) How is the cracking process carried out and what happens to the alkane molecules during this process?
(*2 marks*)

(ii) Give the molecular and structural formula of the alkene, propene.
(*2 marks*)

(iii) Describe a chemical test which can be used to distinguish propane gas from propene gas.
(*3 marks*)

c) Ethanoic acid shows the typical properties of a dilute acid. It also reacts with ethanol to produce the ester, ethyl ethanoate.

(i) Describe what you would observe when solid sodium carbonate is added to dilute ethanoic acid.
(*2 marks*)

(ii) Give one way of making ethanol.
(*1 mark*)

(iii) Describe how you would prepare ethyl ethanoate in the laboratory using ethanoic acid and ethanol.
(*2 marks*)

(iv) Give the full structural formula for ethyl ethanoate. All bonds should be shown.
(*2 marks*)

Chapter 16

Rates of Reaction

Learning objectives

By the end of this chapter you will be able to:

➤ Describe the qualitative effects of temperature, concentration, particle size, catalysts and light on the rate of a chemical reaction.

➤ Give examples of catalysts for specific reactions, for example, MnO_2 with H_2O_2

➤ Understand the use of catalysts in the manufacturing process

➤ Identify the significant factors which control the rates of reaction and their quantitative effects. Quantitative effects are limited to the interpretation of data, for example, drawing graphs and making predictions about how the rate may change when different factors are altered

➤ Give a simple explanation of how the factors affecting the rate of reaction influence the rate in terms of collisions and the energies of the reacting particles

➤ Relate the factors affecting the rate of chemical reaction to the practical problems associated with the manufacturing process and recognise the need for compromise between competing priorities. These are limited to the processes mentioned in the syllabus, for example, ammonia, sulphuric acid and nitric acid

Chemical reactions

We are familiar with many chemical reactions in everyday life that take place at different rates. In the kitchen not all foods cook at the same rate, for example, it would take about ten minutes to fry chips but only a few minutes to toast bread or boil an egg. A slower, less favourable reaction is the souring of milk. This is caused by bacteria and takes a few days to happen. An even slower reaction is the ripening of fruit and this takes place over a period of several weeks.

Some chemical reactions are very slow and it may take many years for us to notice them. Two common examples are the reaction of acid rain with limestone buildings or the effects of rusting on an old car. We can also think of many reactions which are very fast and take place in a fraction of a second, for example baking soda reacts immediately with vinegar giving off bubbles of carbon dioxide gas while explosions in quarries to break rocks seem to occur instantaneously. In the laboratory we can show the presence of hydrogen gas in a test tube by putting a lighted splint to it and observing the immediate small 'pop' or explosion taking place.

Figure 1 Many reactions take place at different rates

While the above examples show that chemical reactions can take place at very different speeds, chemists have found that they can vary the rate of reaction by controlling factors such as:

- temperature
- surface area
- concentration
- catalysts
- light.

In the chemical industry chemists are involved with rates of reaction in the production of many commercial substances. Some examples are the manufacture of ammonia and nitric acid for making fertilisers while sulphuric acid is used in the production of detergents.

Not only is it important for chemists to make these commercial products but it is also essential that products are obtained at a minimum cost and are made as quickly as possible. For industrial chemical reactions it is essential that economic yields can be achieved in as short a time as possible.

A good example of the use of rates of reaction in industry is in the production of ammonia to make fertilisers. At room temperature and pressure nitrogen and hydrogen do not react to produce ammonia; however, if the reaction mixture is heated to 450°C and is submitted to a pressure of 250 atmospheres, in the presence of an iron catalyst, chemists have found that the rate of reaction greatly increases and the production of ammonia is economically worthwhile.

Before studying the factors that affect the rate of reaction in detail it is necessary to look at what is meant by rate of reaction.

Measuring the rate of everyday reactions

Some common examples of rate are:

- a car travelling a distance of 160 km in 2 hours is said to be travelling at a rate of 80 km per hour
- 20 litres of diesel being pumped into a tank in two minutes, the rate is said to be 10 litres per minute
- a factory producing 1000 bottles of sparkling water every 20 minutes, the rate is 50 bottles per minute

From the above three examples it is seen that rate is the change in a measured quantity divided by the time taken:

$$\text{Rate} = \frac{\text{change in a measured quantity}}{\text{time taken}}$$

Thus for a car travelling 160 km in two hours the rate is obtained by dividing 160 km by 2 and this gives a value of 80 km per hour. In a chemical reaction the rate is written as:

$$\text{Rate} = \frac{\text{change in amount of reactant or product}}{\text{time taken}}$$

Measuring the rate of reaction in a chemical reaction

The rate of reaction tells us how fast or how slowly a chemical reaction is taking place. Take, for example, the reaction between magnesium and dilute sulphuric acid producing hydrogen gas and magnesium sulphate.

Magnesium + sulphuric acid $\rightarrow$ magnesium sulphate + hydrogen
$$Mg(s) \quad + \quad H_2SO_4(aq) \quad \rightarrow \quad MgSO_4(aq) \quad + \quad H_2(g)$$

As the reaction progresses the rate of reaction can be measured in terms of:

a) **the amount of reactant used up in a given time**
In the above reaction this could be the amount of magnesium or sulphuric acid used up per minute; or

b) **the amount of product obtained in a given time**
For the same reaction this could be the amount of magnesium sulphate or hydrogen formed per minute.

Thus if 0.1 gram of magnesium was added to dilute sulphuric acid and it took 20 seconds for the magnesium to react completely and give 100 cm^3 hydrogen gas, the rate of reaction can be expressed as:

$$\text{Rate} = \frac{\text{change in mass of magnesium or the volume of hydrogen given off}}{\text{time taken}}$$

$$= \frac{0.1}{20} = 0.005 \text{ grams per second (gs}^{-1}) \text{ or } \frac{100}{20} = 5 \text{ cm}^3 \text{ per second (cm}^3\text{s}^{-1})$$

The rate calculated above is the **average rate** over 20 seconds for all the magnesium to react or for the total volume of hydrogen to form.

Factors affecting the rate of reaction

In this section the factors which affect the rate of reaction will be studied.

Surface area

Consider the following chemical reactions:

- cooking: potatoes take a shorter time to cook when they are cut into smaller pieces
- combustion: it is easier to light a fire with small pieces of stick, rather than by using large blocks of wood
- combustion: some modern coal fired power stations burn powdered coal rather than lumps of coal
- metals reacting with acids: zinc powder will react much more quickly with dilute sulphuric acid than do zinc granules

All of the above reactions occur more quickly because the surface area of the solid has been increased. By surface area we mean the amount of solid surface that is available for reaction. The smaller the pieces of solid then the greater the surface area.

A study of the reaction between marble chips and dilute hydrochloric acid shows that when the surface area of a solid is increased so too is the rate of reaction. Marble chips or calcium carbonate reacts with dilute hydrochloric acid to produce calcium chloride, water and carbon dioxide:

$$CaCO_3(s) + 2HCl(aq) \rightarrow CaCl_2(aq) + H_2O(l) + CO_2(g)$$

The rate of reaction can be studied using the apparatus shown in Figure 2.

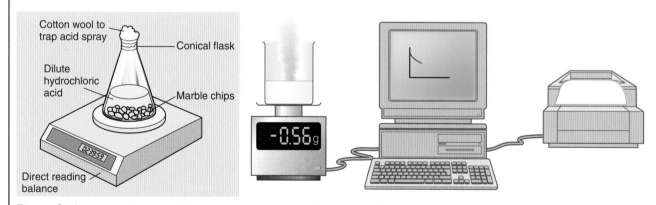

Figure 2 Measuring the rate of reaction between marble chips and hydrochloric acid

In this experiment a given mass of large marble chips (excess) is added to a specified volume of dilute hydrochloric acid. As carbon dioxide gas escapes the mass of the flask and its contents decreases. The loss in mass of the flask and its contents are measured and recorded each minute until the reaction is over. The experiment is repeated again using the same volume and concentration of hydrochloric acid and the same temperature. The same mass of marble chips is also used but this time medium sized chips are used. A third experiment using exactly the same conditions but small marble chips is also carried out.

The results for the three different sizes of marble chips are given in Table 1 while Figure 3 shows graphs for the three sets of results.

Table 1 Results for the three sizes of marble chips

Time (mins)	Total mass loss (g)		
	Large pieces	**Medium pieces**	**Small pieces**
0	0.00	0.00	0.00
1	1.00	1.48	2.96
2	1.80	2.52	3.68
3	2.52	3.20	3.85
4	3.00	3.58	3.93
5	3.29	3.73	3.98
6	3.50	3.85	3.99
7	3.66	3.92	4.00
8	3.76	3.96	4.00
9	3.82	3.99	4.00
10	3.88	4.00	4.00
11	3.93	4.00	4.00
12	3.97	4.00	4.00
13	3.99	4.00	4.00
14	4.00	4.00	4.00
15	4.00	4.00	4.00

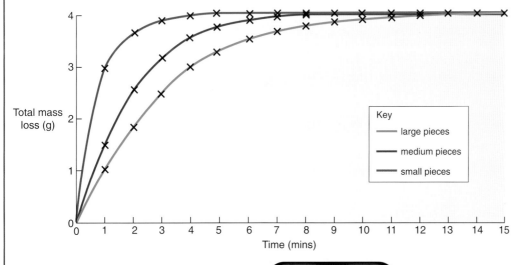

Figure 3 How the surface area of marble chips affects the rate of reaction with hydrochloric acid

Key
— large pieces
— medium pieces
— small pieces

Questions

1   This question refers to the results given in Table 1 and Figure 3.

a) Look at the results in Table 1 and calculate in which reaction there was the greatest loss in mass after 5 minutes.

b) Calculate the average rate for each reaction after 5 minutes and work out in which reaction the rate is fastest.

c) From the graphs in Figure 3 decide which reaction is taking place most quickly after 5 minutes? Explain your answer.

d) Look closely at the results and decide which reaction finishes first. Write down how long it takes for each reaction to reach completion.

e) What mass of carbon dioxide is formed at the end of each reaction?

f) For each graph what happens to the reaction rate as time progresses?

The above results show that for each experiment there is the same loss in mass of carbon dioxide in each experiment. The slopes of the graphs also show the same pattern in that each time the reaction rate starts off fast, then slows down and eventually stops. Most importantly, the graphs demonstrate that the rate of reaction is fastest when small marble chips are used. The steeper the slope or the greater the gradient then the faster the reaction is taking place.

Explaining the effect of surface area on reaction rate

Collision theory is used to explain how different factors affect the rate of a chemical reaction. The collision theory states that for a reaction to take place between reacting particles (ions, molecules or atoms) it is necessary that they collide. Additionally, particles must have sufficient energy to react, otherwise they simply bounce off each other without reacting. This minimum amount of energy that reacting particles must possess to react is called the **activation energy** and this energy is used to break bonds in the reacting particles in order that products can form.

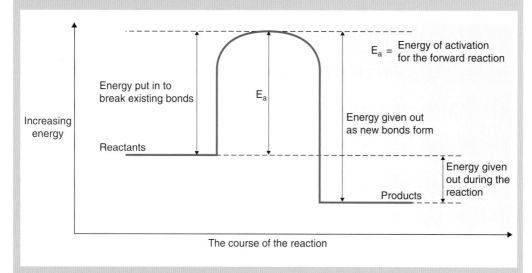

Figure 4 Activation energy for a chemical reaction

In the reaction between marble chips and hydrochloric acid, it is the hydrogen ions of the hydrochloric acid that react with the carbonate particles on the surface of the marble chips. The hydrogen ions are moving freely throughout the acid solution and a reaction can take place, when hydrogen ions with sufficient energy collide with the carbonate particles. As a result of these successful collisions, bubbles of carbon dioxide are produced:

$$CO_3^{2-} + 2H^+ \rightarrow CO_2 + H_2O$$

Each time there is a successful collision the hydrogen ions are used up. Due to the decreasing concentration of hydrogen ions as time progresses the rate of reaction decreases. If the marble chips are in excess then the reaction ends when all the hydrogen ions are used up and no more bubbles of carbon dioxide will be given off.

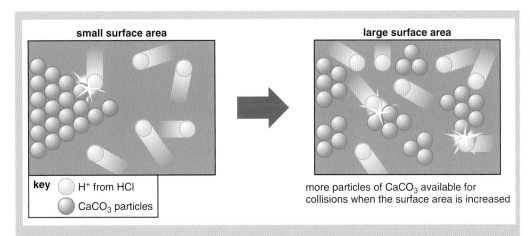

small surface area   large surface area

key H^+ from HCl

$CaCO_3$ particles

more particles of $CaCO_3$ available for collisions when the surface area is increased

Figure 5 Marble chips reacting with hydrochloric acid

When the marble chips in the above reaction are broken up into smaller pieces they have a greater surface area as shown in Figure 5. This means that there will be more particles available for reaction. In a given time there will be more successful collisions between the hydrogen ions and the carbonate particles of the marble and as a result the rate of reaction will be faster.

Concentration

The reaction between sodium thiosulphate and dilute hydrochloric acid can be used to demonstrate how the concentration of a reactant can affect the rate of a chemical reaction.

$$\begin{array}{c} \text{Sodium} \\ \text{thiosulphate} \end{array} + \begin{array}{c} \text{hydrochloric} \\ \text{acid} \end{array} \rightarrow \begin{array}{c} \text{sodium} \\ \text{chloride} \end{array} + \begin{array}{c} \text{sulphur} \\ \text{dioxide} \end{array} + \text{sulphur} + \text{water}$$

$$Na_2S_2O_3(aq) + 2HCl(aq) \rightarrow 2NaCl(aq) + SO_2(g) + S(s) + H_2O(l)$$

In this reaction the concentration of sodium thiosulphate (see Figure 6) is varied and the time for a given mass of sulphur to form recorded. The reacting solutions are mixed and placed on top of a filter paper with a cross drawn on it and a stop clock is started just at the same time. As the reaction proceeds sulphur starts to form and the reaction mixture begins to go cloudy. When the cross on the filter paper is just no longer visible the clock is stopped and the time recorded for the given mass of sulphur to form. The reaction is repeated for a number of different concentrations of sodium thiosulphate.

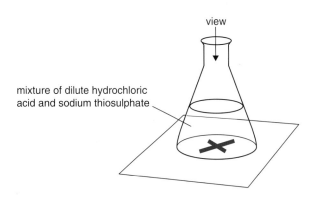

view

mixture of dilute hydrochloric acid and sodium thiosulphate

Figure 6 Reaction between sodium thiosulphate and hydrochloric acid

Table 2 Typical results for the sodium thiosulphate and hydrochloric acid reaction

Experiment number	Volume of sodium thiosulphate used (cm³)	Volume of water used (cm³)	Volume of Hydrochloric acid used (cm³)	Time (sec)	1/time (sec⁻¹)
	Concentration				Rate
1	50	0	5	38	0.0263
2	40	10	5	47	0.0213
3	30	20	5	62	0.0161
4	25	25	5	74	0.0135
5	20	35	5	95	0.0105
6	10	40	5	182	0.0055

Questions

2 This question relates to the results in Figure 2 for sodium thiosulphate reacting with hydrochloric acid.

a) How was the concentration of sodium thiosulphate changed for each reaction mixture in the experiment?

b) For each reaction at what stage is the clock stopped?

c) How does the concentration of sodium thiosulphate affect the time for the reaction?

d) Explain why the rate of reaction changes when the concentration of a reactant changes.

From the results in Table 2 it is seen that as the concentration of the thiosulphate decreases then the time for the reaction increases and the rate of reaction gets slower. The rate is inversely proportional to the time taken and we can write:

rate of reaction $\propto$ 1/time (note: 1/ time can be used as a measure of rate)

Figure 7a shows the graph produced when the volume of sodium thiosulphate (concentration) is plotted against time for the reaction and in 7b when it is plotted against 1/time.

The graphs show how the concentration of sodium thiosulphate affects the time for the reaction to take place and also how the concentration of sodium thiosulphate affects the rate of reaction. As the thiosulphate concentration is increased then the time for the reaction decreases or the rate of reaction increases. Collision theory can be used to explain why the rate of reaction increases when the concentration of a reactant is increased. Figure 8 shows that when the concentration of sodium thiosulphate is low, there are only a small number of collisions with the H^+ ions of hydrochloric acid in a given time. Considering that only some of the collisions have enough energy for a reaction to occur it is not surprising that the rate of reaction is slow. As the concentration of thiosulphate increases there are more particles available to collide with the H^+ ions and so there are more successful collisions in a given time and the rate of reaction is faster.

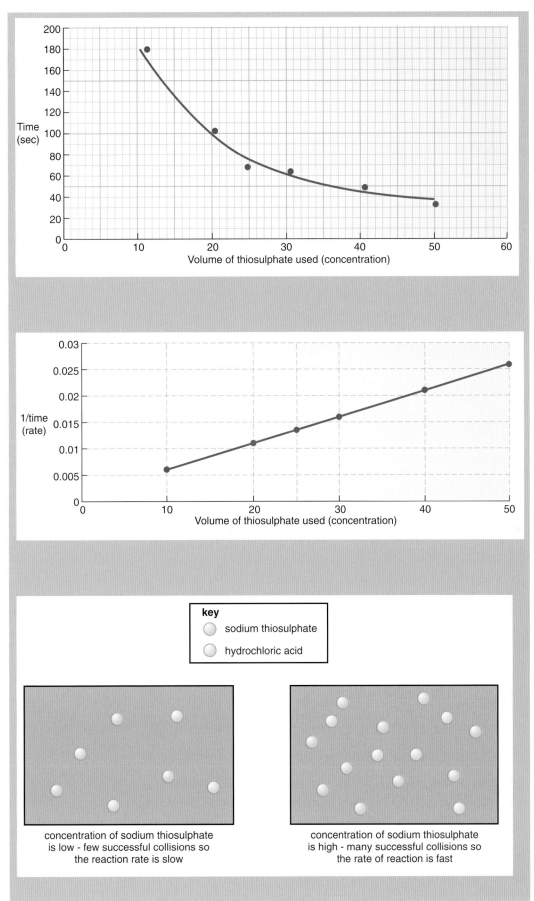

Figure 7a Volume of sodium thiosulphate (concentration) against time for a reaction

Figure 7b Volume of sodium thiosulphate (concentration) against 1/time or rate of reaction

Figure 8 Using collision theory to explain the effect of concentration on the rate of reaction

Temperature

The effect of temperature on the rate of a chemical reaction can be studied for the sodium thiosulphate solution and dilute hydrochloric acid reaction. The reaction is carried out as in the previous section but for only one concentration of sodium thiosulphate solution at a number of different temperatures. The cross is viewed as before and for each experiment the time for it to just disappear is noted. Table 3 gives a typical set of results.

Table 3 Table of results to show the effect of temperature on the rate of reaction

Temperature (°C)	Time for cross to disappear (s)	$\dfrac{[1]}{\text{time for cross to disappear}} (s^{-1})$
23	132	0.0076
29	90	0.0111
34	65	0.0154
39	46	0.0217
44	33	0.0313

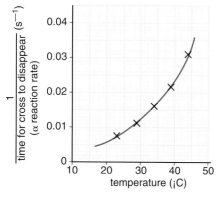

Figure 9 Graph to show the effect of temperature on the rate of reaction

From the table of results and the graph in Figure 9 it is seen that as the temperature is increased the time for the reaction is reduced and the rate of reaction is increased.

In a chemical reaction when the temperature is increased the reacting particles gain energy and move faster. As the particles are moving faster there are more collisions in a given time. Since collisions are more energetic, it is not surprising that when the temperature is increased the rate of reaction is markedly increased. For most reactions the rate is approximately doubled when the temperature is increased by only 10°C. The reason for this is that a rise of 10 degrees approximately doubles the number of particles with energy equal to the activation energy. Thus when the temperature is raised by 10 degrees there are twice as many successful collisions in a given time and the rate is approximately doubled.

Questions

3 Magnesium carbonate reacts with dilute hydrochloric acid as shown below.

$$MgCO_3(s) + 2HCl(aq) \rightarrow MgCl_2(aq) + H_2O(l) + CO_2(g)$$

A student decided to investigate the rate of this reaction at different temperatures. She carried out the reaction in a flask placed on a balance and measured the loss in mass at different times. Her results are shown below.

Time (min)	0	1	2	3	4	5	6
Total mass (g)	100	99.2	98.5	98	97.7	97.5	97.5

a) Draw a diagram of the apparatus showing how the loss in mass can be calculated.
b) Plot the results on graph paper and draw a line of best fit.
c) Explain why there is a loss in mass as the reaction proceeds.
d) Describe how the rate of reaction changes with time.
e) Draw a curve on the graph to show what would happen over the first six minutes, if the reaction was repeated exactly as before but with the temperature raised by about 10°C. Label this curve P.
f) Explain, in terms of particles, why raising the temperature of the acid affects the rate of reaction.

Catalysts

There are many chemical reactions that are speeded up by using a **catalyst**. A catalyst is a substance which speeds up the rate of a chemical reaction without itself undergoing any permanent chemical change.

As catalysts are not being used up during chemical reactions, it means that only small amounts of catalysts are required to speed up reactions. A small amount of catalyst is capable of speeding up the conversion of an infinite amount of reactant to product.

Consider the decomposition of hydrogen peroxide solution using the catalyst manganese(IV) oxide. At room temperature and without a catalyst this reaction is very slow. However, when a small amount of solid manganese(IV) oxide is added the decomposition is very fast and oxygen is given off rapidly.

$$H_2O_2(aq) \rightarrow 2H_2O(l) + O_2(g)$$

Figure 10 shows the effect of changing the amount of manganese(IV) oxide on the time to produce 50 cm^3 oxygen gas from hydrogen peroxide.

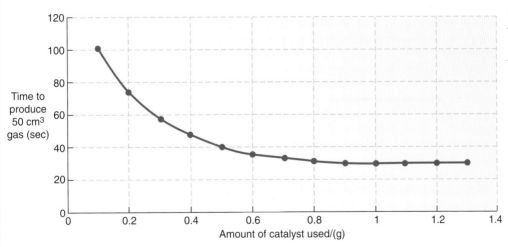

Figure 10 The effect of the amount of manganese dioxide on the rate of reaction

From Figure 10 it is seen that the rate of reaction increases as the mass of manganese(IV) oxide increases. This occurs up to 0.9 g but after this value the rate does not increase when the amount of catalyst is increased.

Why catalysts speed up the rate of reaction

For a reaction to take place particles must collide and they must have sufficient energy to overcome the activation energy. If not, they merely collide and bounce off each other. Catalysts increase the rate of reaction by lowering the activation energy of a reaction. This means that in a catalytic reaction when particles collide more of them have sufficient energy to overcome the energy barrier to form products and the reaction is faster. This is shown in Figure 11

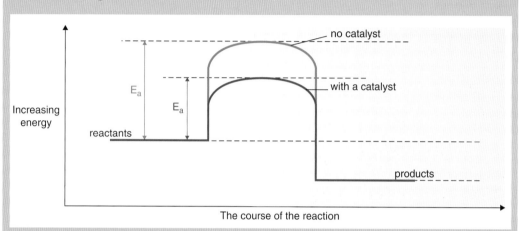

Figure 11 An energy level diagram showing why a catalyst speeds up a reaction

Some important examples of catalysts

Contact process to make sulphuric acid

Vanadium pentoxide, V_2O_5, is used as a catalyst to convert sulphur dioxide to sulphur trioxide at a temperature of 450°C. Sulphur trioxide is then converted to sulphuric acid. Sulphuric acid is used in car batteries and in the manufacture of detergents, fibres, and pigments.

$$\text{sulphur dioxide} + \text{oxygen} \rightleftharpoons \text{sulphur trioxide}$$
$$2SO_2(g) + O_2(g) \rightleftharpoons 2SO_3(g)$$

The production of nitric acid by the catalytic oxidation of ammonia

In the first stage of the production of nitric acid a platinum catalyst at 900°C is used to convert ammonia into nitrogen monoxide.

$$\text{ammonia} + \text{oxygen} \rightleftharpoons \text{nitrogen monoxide} + \text{water}$$
$$4NH_3(g) + 5O_2(g) \rightleftharpoons 4NO(g) + 6H_2O(g)$$

Catalytic cracking

In oil refineries catalysts are used to crack large hydrocarbons molecules into smaller more useful ones. A zeolite catalyst at 500°C is used.

$$\text{decane} \rightarrow \text{octane} + \text{ethene}$$
$$C_{10}H_{22} \rightarrow C_8H_{18} + C_2H_4$$

Ethene can be used to make polythene while octane is used as petrol for cars.

Fermentation

In brewing during the fermentation process enzymes in yeast break down sugars to form ethanol and carbon dioxide.

$$2C_6H_{12}O_6 \rightarrow 2C_2H_5OH + 2CO_2$$
$$\text{glucose} \quad\quad \text{ethanol} \quad\quad \text{carbon dioxide}$$

Biological washing powders

Biological washing powders that are effective in removing stains in the form of food and blood from clothes contain the enzymes lipases and proteases. These enzymes break down stains caused by fats and proteins and have the advantage of working at a low temperature of 40°C.

Figure 12 A biological washing powder

Questions

4 Manganese(IV) oxide catalytically decomposes hydrogen peroxide into water and oxygen.

a) What is the meaning of the term catalyst?
b) Write a balanced equation to show the decomposition of hydrogen peroxide.
c) In the above reaction the mass of manganese(IV) oxide remains constant throughout the reaction. Describe an experiment you could carry out to show that the catalyst, manganese(IV) oxide is not used up during the reaction.

Light

Light is a form of energy and it causes many chemical reactions to take place. Some photochemical reactions need the continual presence of light while others only require light to initiate them. Examples of photochemical reactions are photosynthesis, hydrogen reacting with chlorine (this reaction is explosive in sunlight) and in photography where light reduces silver ions to metallic silver in the photographic film. During the developing stage of the film more silver is deposited on the existing silver and a negative is formed.

Rates of reaction in industry

For the production of any material or product chemists want to obtain the maximum amount of material in the shortest time possible and as cheaply as possible. To achieve this they must they must:

- use cheap reactants
- use a fast reaction rate
- obtain the maximum possible yield of product
- use the minimum amount of energy to obtain the product.

Looking at these factors we can see how they apply to the production of ammonia in the Haber–Bosch process. Ammonia is an important commercial chemical because it is used to produce fertilisers, nylon and nitric acid. Figure 6, Chapter 13 shows how ammonia is produced from air, naptha and natural gas in the Haber–Bosch Process.

The reaction between nitrogen and hydrogen is an equilibrium reaction and is carried out using the following conditions:

● an iron catalyst
● a temperature of 450°C
● a pressure of 250 atmospheres.

These conditions are a compromise so that the equilibrium reaction is pushed to the right hand side to:

● give a reasonable yield of ammonia
● ensure that the reaction rate is sufficiently fast to give an economic conversion of nitrogen and hydrogen to ammonia.

$$\text{nitrogen} + \text{hydrogen} \rightleftharpoons \text{ammonia}$$
$$N_2(g) + 3H_2(g) \rightleftharpoons 2NH_3(g)$$

Using the above conditions almost 25 percent of the hydrogen and nitrogen mixture is converted to ammonia. To ensure that no reactants are wasted, the hot gaseous mixture is cooled until ammonia liquifies ($-33°C$). This allows it to be separated from the hydrogen and nitrogen, which are then recycled to produce more ammonia.

High pressure pushes the equilibrium to the right hand side and this gives a higher yield of ammonia. In any gaseous equilibrium, increasing the pressure shifts the equilibrium to the side that has fewer molecules. In the formation of ammonia, three molecules of hydrogen and one molecule of nitrogen react to form two molecules of ammonia, thus when the pressure is increased the equilibrium shifts to favour the production of ammonia. This means that more ammonia forms at equilibrium. Very high yields of ammonia can be achieved using pressures of up to 600 atmospheres; however, it is expensive both to make and to maintain reacting vessels operating at such high pressures.

The reaction of nitrogen and hydrogen is an exothermic reaction because heat energy is produced. For an exothermic equilibrium a low temperature would theoretically produce the highest yield of product (ammonia in this case). At first we might think that it would be more suitable to use a low temperature because it would eliminate the high cost of electricity and at the same time give a high yield of ammonia. However, at low temperatures the rate of reaction is so slow that the process would be uneconomical and so higher temperatures are used. Figure 13 shows how the percentage yield of ammonia changes according to temperature and pressure.

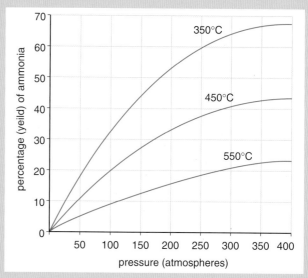

Figure 13 The percentage yield of ammonia at equilibrium under different temperatures and pressures

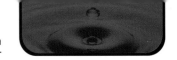

Under the operating conditions of 450°C and 250 atmospheres, the iron catalyst lasts about five years. As the iron catalyst plays an important part in speeding up the rate of attainment of the equilibrium, it is important that the incoming nitrogen and hydrogen are purified before the reaction takes place. Impurities such as carbon monoxide destroy or poison the catalyst.

Without an iron catalyst the process would be impossible to carry out. The iron catalyst not only decreases the time for the equilibrium to be attained but it also allows the reaction to take place at lower temperature and pressure and this makes the production of ammonia more economical.

The cost of the raw materials is kept to a minimum, nitrogen is obtained by the fractional distillation of liquid air while hydrogen is readily obtained from naptha or natural gas. Air is free while naptha and natural gas are relatively cheap and obtained from petrochemicals. Iron which is used as the catalyst is easy and cheap to manufacture from haematite. Nevertheless, there are high energy costs involved with several stages in the process:

- the high reaction temperature of 450°C
- the conversion of naptha or methane to hydrogen
- pumping the gases to the catalytic convertor and also in recycling hydrogen and nitrogen back to the convertor.

The Haber–Bosch process highlights the need for industrial chemists to consider rates of reactions in the manufacture of commercial products. It shows why decisions have to be taken on the need to compromise between factors such as the cost of plant equipment, reaction conditions and product yield, in order that products such as ammonia can be produced as economically as possible.

Questions

5 a) List the factors that can affect the rate of a chemical reaction
   b) Name the catalyst which is used in
      (i)   the Haber–Bosch Process
      (ii) the Contact Process
      (iii) the production of nitric acid by the catalytic oxidation of ammonia.
   c) Why are catalysts important in industry?

6 Ammonia is produced in the Haber–Bosch Process.
   a) Write an equation to show how hydrogen reacts with nitrogen to form ammonia.
   b) What type of reaction is this?
   c) Give the conditions which are used in the Haber–Bosch Process and explain why these conditions are used.
   d) Why is this reaction not demonstrated in school laboratories?

7 Use collision theory and activation energy to explain how a catalyst works.

8 Use the website http://www.creative-chemistry.org.uk/gcse/module7.htm and answer the questions on the production of ammonia on the worksheet 'Making Ammonia'.

Exam questions

1 A group of students investigated how 25 cm³ of hydrogen peroxide solution breaks down to produce oxygen. The volume of oxygen collected was measured every minute and a graph drawn as shown below.

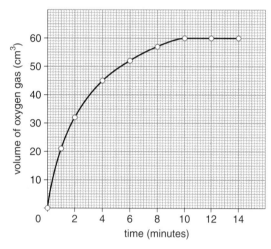

a) Name a piece of apparatus which would be suitable to collect the oxygen gas. *(1 mark)*

b) The students repeated the experiment but the only change was to add 10 cm³ water to 25 cm³ of the hydrogen peroxide solution. On the same grid sketch the curve you would expect for this reaction and label it with the letter B. *(2 marks)*

c) Why might a catalyst be added to the hydrogen peroxide solution? *(1 mark)*

d) In terms of **particles** explain why the rate of a chemical reaction increases when the temperature is increased.

(2 marks)

2 Carbon dioxide gas is often made in laboratories by reacting calcium carbonate with hydrochloric acid.

$$CaCO_3 + 2HCl \rightarrow CaCl_2 + H_2O + CO_2$$

A chemist carried out five experiments to study this reaction. Each experiment used the same mass of calcium carbonate and the same volume of acid.

Experiment	Description of calcium carbonate	Temperature °C	Concentration of acid in mol/dm³
A	lump	20	0.1
B	powder	20	0.2
C	lump	40	0.1
D	powder	40	0.1
E	powder	40	0.2

a) Which experiment would have the fastest reaction? *(1 mark)*

b) In terms of how particles react, explain why this experiment was the fastest. *(3 marks)*

3 A student investigated the reaction between 0.09 g of magnesium ribbon and excess dilute sulphuric acid at 30°C. The gas was collected and its volume measured every 20 seconds.

The following results were obtained.

Time (s)	0	20	40	60	80	100	120	140	160
Vol (cm³)	0	30	48	64	74	82	88	90	90

a) Copy and complete the diagram to show how the gas could be collected and its volume measured.

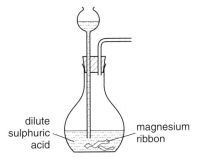

dilute sulphuric acid

magnesium ribbon

(2 marks)

b) Draw a graph of the results.

(3 marks)

c) How long did it take to collect 60 cm³ of gas?

(1 mark)

d) The student repeated the experiment but the only change made was to reduce the temperature to 20°C. On the **same** grid sketch the curve you would expect from this investigation and label it with the letter S.

(2 marks)

e) In terms of particles explain how the rate of reaction is affected by decreasing the temperature.

(3 marks)

4 a) As part of an investigation to show how the rate of reaction is affected by particle size, a student set up the following apparatus.

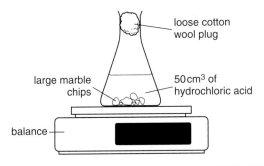

loose cotton wool plug

large marble chips

50 cm³ of hydrochloric acid

balance

The mass of the flask and its contents is recorded at 30 second intervals after the addition of the acid until the mass becomes constant. Some of the marble chips remained unreacted at the end of the experiment.

(i) Write a **balanced symbol** equation for the reaction which occurs during the experiment.

(2 marks)

(ii) Explain why the mass of the flask and its contents **decreases** during the experiment.

(2 marks)

b) The results obtained during one such experiment are shown in the table below.

Time (s)	0	30	60	90	120	150	180	210	240	270
Mass lost (g)	0	0.54	0.82	1.0	1.12	1.14	1.26	1.28	1.30	1.30

(i) Plot these results on a graph and use it to answer the following questions.

(5 marks)

(ii) Which point on the graph is clearly wrong and should be checked if possible.

(1 mark)

(iii) Over which time interval is the rate of reaction greatest? Explain your answer.

(3 marks)

(iv) The experiment was repeated using **smaller** marble chips of the same total mass and using the same volume of acid of the same concentration. Sketch on your graph a curve which would show how they would react. Label this curve A.

(3 marks)

(v) Name **two** other factors, apart from particle size, which could alter the rate of this reaction and describe how the rate is altered in each case. (Assume that no catalyst is available for this reaction.)

(6 marks)

(vi) Using **one** of the factors listed in part (v), explain, using collision theory how this alters the rate of reaction.

(3 marks)

Radioactivity

By the end of this chapter you will:

➤ Know what is meant by ionising radiation

➤ Find out about the different types

➤ Know about radioactive decay, alpha and beta particles and gamma radiation

➤ Know what happens inside the nucleus when a radioactive isotope decays and how to write equations for the processes

➤ Understand what is meant by 'half-life' and be able to carry out simple calculations involving half-lives

Note: Chapter 17 is for Double Award Science students only.

Radioactive material is found naturally all around us and inside our bodies. A small number of carbon atoms are radioactive carbon-14 isotopes, they are found in the carbon dioxide in the air and in the cells of all living organisms. Traces of radioactive elements, for example potassium, can be found in our food. Certain rocks contain uranium, all the isotopes of which are radioactive, and this decays giving radon, a radioactive gas. There is also radiation reaching the Earth from outer space. All these natural sources are known together as background radiation.

Among the scientists who carried out early work on radioactivity are Henri Becquerel and Marie and Pierre Curie. They were jointly awarded the Nobel Prize for Physics in 1903.

Did you know?

Pierre and Marie Curie worked with 8 tonnes of pitchblende and only obtained 1 g of radium chloride from it.

Ionising radiation

The nuclei of some atoms are unstable and they emit radiation. This is known as **ionising radiation** because while it passes through matter it causes some of the atoms to become ions.

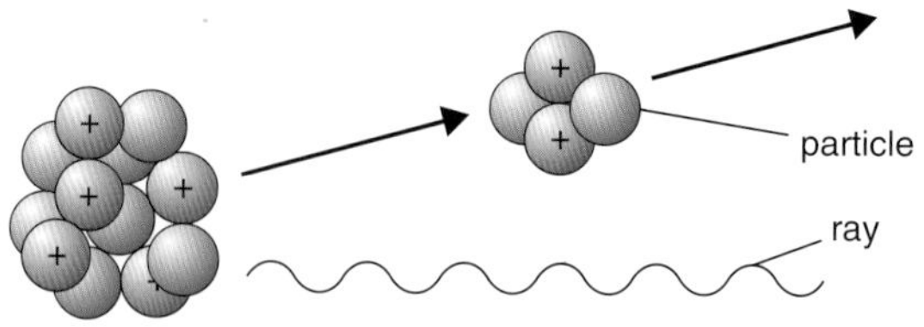

Types of radiation

Alpha (α) radiation

Figure 1 Unstable nucleus emitting a particle and a ray

● It is a stream of **alpha particles** emitted from large nuclei.
● An alpha particle is a helium nucleus, i.e. it is two protons plus two neutrons and so it has a relative atomic mass of 4.
● An alpha particle is positively charged.
● An alpha particle will be deflected in a magnetic field.
● Alpha particles have poor powers of penetration and can only travel through a few centimetres of air.

- A sheet of paper is enough to stop alpha particles.
- Alpha radiation has the strongest ionising power.
- The uranium-238 isotope decays by emitting alpha radiation.

Beta (β) radiation

- It is a stream of **beta particles** being emitted from nuclei where the number of neutrons is much larger than the number of protons.
- A beta particle is an electron which has been formed in the nucleus and thus it has relative atomic mass of about $\frac{1}{2000}$.
- As a beta particle is negatively charged it will be deflected in a magnetic field and because it is much smaller than an alpha particle the deflection will be greater and in the opposite direction.
- Beta particles move much faster than do alpha particles and so have a greater penetrating power.
- A beta particle can travel for 50 cm in air.
- A beta particle is stopped by 0.5 cm thick aluminium foil.
- Beta radiation has an ionising power between that of alpha and gamma radiation.
- The carbon-14 isotope decays by emitting beta radiation.

Gamma (γ) radiation

- Unlike the other types it does not consist of particles but of high energy waves.
- Because there are no particles gamma radiation has no mass.
- Because there are no charged particles a magnetic field has no effect on gamma radiation.
- It has great penetrating power, travelling several metres in air.
- A thick block of lead or concrete is used to stop gamma radiation and in fact this does not completely halt it but its effects are greatly reduced.
- It has the weakest ionising power.

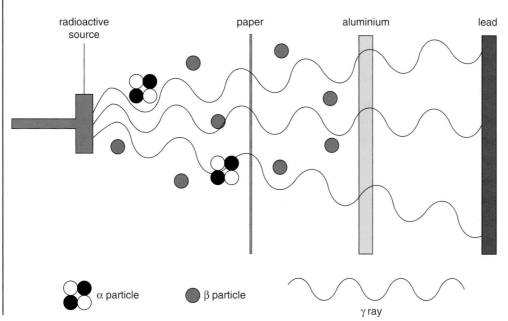

Figure 2 Selective absorption of radioactive emissions

Uses of radiation

● Carbon dating of organic material. All living organisms contain some carbon-14 When the organism is alive the ratio of carbon-12 to carbon-14 remains constant. Once the organism dies the amount of carbon-14 decreases as the radioactive isotope decays. Comparing the amount of carbon-14 present in a sample with the amount in a living organism allows calculation of the age of the sample. Fortunately carbon-14 has a long half-life and so it decays slowly and there is enough left to measure. This method was used to date the Dead Sea Scrolls.

● Gamma radiation from the cobalt-60 isotope can be used to treat tumours.

● Gamma radiation can be used to treat fresh food. It kills bacteria on the food and therefore the food will keep for longer. The use, however, is controversial as many people are worried about the long terms effects on the human body of eating irradiated food.

● Surgical instruments and hospital dressings can also be sterilised by exposure to gamma radiation.

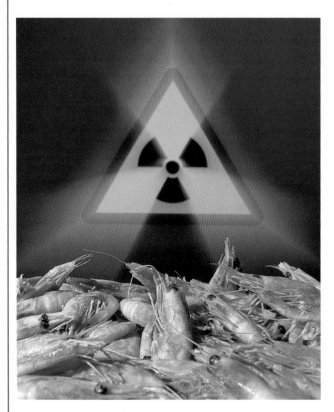

Figure 3a Harmful bacteria in food can be killed by gamma rays

Figure 3b Operating equipment is sterilised by gamma rays

- The thickness of a sheet of paper or aluminium can be checked and possibly controlled using beta radiation. An emitter is on one side of the sheet and a detector on the other. As the sheet moves past the activity detected will be the same as long as the thickness remains unchanged.

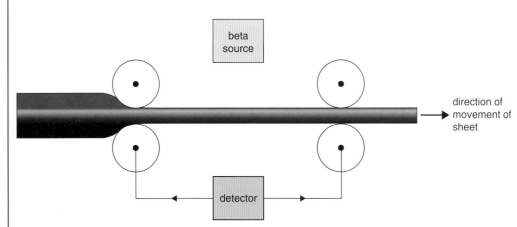

- Information about fluid movement and mixing can be obtained using a suitable radioactive isotope, for example leaks in underground pipes.

- Iodine-131 is used in investigations of the thyroid gland.

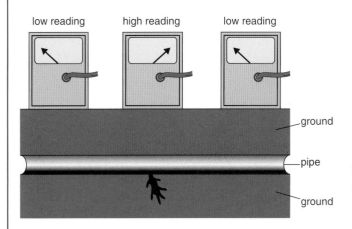

Figure 5 Using radioactive tracers to locate a leak in a pipe

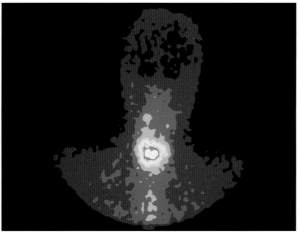

Figure 6 Iodine-131 localised in the thyroid gland

Care must be taken when using radioactive isotopes because the radiation can damage living cells by altering the structure of the cell's chemicals. Protective clothing must be worn and the amount of time that the worker is exposed to the radiation must be controlled. Any radioactive isotopes that are taken internally are not usually alpha emitters, (they are too powerful ionisers), and they need to have short half lives so that they do not remain for too long in the tissues.

Nuclear equations

Symbol equations can be written to represent alpha and beta decay. The alpha particle can be written as α or as $^4_2$He and the beta particle as β or $^0_{-1}$e.

Examples:

1 Alpha decay of uranium-238

$$^{238}_{92}\text{U} \rightarrow \,^{234}_{90}\text{Th} + \,^4_2\text{He (or } \alpha)$$

2 Beta decay of carbon-14

$$^{14}_{6}\text{C} \rightarrow \,^{14}_{7}\text{N} + \,^0_{-1}\text{e (or } \beta)$$

You need to remember the following:

● The total mass number on the left hand side must equal the total mass number on the right hand side.

● The total atomic number on one side must equal the total atomic number on the other side.

If you know the original isotope and the one formed by the decay then it is possible to determine the type of decay by working out the type of particle emitted.

If you know the original isotope and the type of decay you can work out the isotope that is formed by the decay.

Examples:

1 Radium-226 decays to polonium-222. Which type of decay occurs?

$$^{226}_{86}\text{Ra} \quad \rightarrow \quad ^{222}_{84}\text{Po} \quad + \,^a_b\text{X}$$

mass number: $\quad 226 \;=\; 222 + a$

$\qquad\qquad\qquad\;\; a \;=\; 226 - 222$

$\qquad\qquad\qquad\;\; a \;=\; 4$

atomic number $\;\;\; 86 \;=\; 84 + b$

$\qquad\qquad\qquad\;\; b \;=\; 86 - 84$

$\qquad\qquad\qquad\;\; b \;=\; 2$

The particle with a mass number of 4 and an atomic number of 2 is helium and so X is an alpha particle. This type of decay is alpha decay.

2 Which isotope is formed by the beta decay of thorium-234?

$$^{234}_{90}\text{Th} \quad \rightarrow \quad ^c_d\text{Y} \quad + \,^0_{-1}\text{e}$$

mass number: $\quad 234 \;=\; c + 0$

$\qquad\qquad\qquad\;\; c \;=\; 234 - 0$

$\qquad\qquad\qquad\;\; c \;=\; 234$

atomic number: $\quad 90 \;=\; d + (-1)$

$\qquad\qquad\qquad\;\; d \;=\; 90 - (-1)$

$\qquad\qquad\qquad\;\; d \;=\; 91$

The element with atomic number 91 is protactinium and the isotope formed is protactinium-234.

Questions

1 Work out the type of decay in each of the following:
  a) Bismuth-213 to polonium-213
  b) Radium-226 to radon-222
  c) Francium-221 to actinium-217

2 Work the name and mass number of the isotope formed in each of the following.
  a) Alpha decay of polonium-214
  b) Beta decay of lead-212
  c) Beta decay of thallium-210

3 a) How does the value of the mass number change in alpha decay?
  b) How does the value of the atomic number change in alpha decay?
  c) How does the value of the mass number change in beta decay?
  d) How does the value of the atomic number change in beta decay?

Changes in the nucleus

Alpha and beta particles are emitted from the nucleus, the electrons in the shells around the nucleus are not involved in radiation.

Alpha decay: the two protons and two neutrons that make up an alpha particle leave the nucleus together as a helium ion.

Beta decay: the electron that is the beta particle is formed in the nucleus. A neutron changes into a proton, which stays in the nucleus, and an electron, which is emitted.

$$^{1}_{0}n \rightarrow ^{1}_{1}p + ^{0}_{-1}e$$

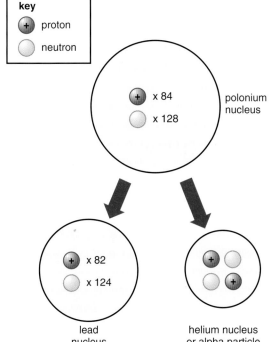

key
+ proton
◯ neutron

polonium nucleus
+ x 84
◯ x 128

lead nucleus
+ x 82
◯ x 124

helium nucleus or alpha particle

carbon-14 nucleus

one neutron becomes a proton and an electron

nitrogen-14 nucleus

e⁻

e⁻

Figure 7 Alpha decay   **Figure 8** Beta decay

Nuclear fission

Nuclear fission (splitting of a nucleus) happens with the release of large amounts of energy.

When uranium-235 is bombarded with fast moving neutrons the uranium nucleus is split into smaller nuclei, strontium-90 and xenon-143. Three more neutrons are produced and an enormous amount of energy is released.

$$^{235}_{92}U + {}^{1}_{0}n \rightarrow {}^{90}_{38}Sr + {}^{143}_{54}Xe + 3\,{}^{1}_{0}n$$

Half-life

As a radioactive isotope decays the activity of the sample decreases. The **half-life** of an isotope is the time taken for the activity to fall by half. An alternative definition states that it is the time taken for the mass of the isotope to fall by half.

Each isotope has a specific and constant half-life. Some half-lives are very short, a matter of seconds or even a fraction of a second, and others can be thousands of years.

Table 1 Half-life data

Isotope	Half-life
uranium-238	4500 000 000 years
carbon-14	5730 years
phosphorus-30	2.5 minutes
oxygen-15	2.06 minutes
barium-144	11.4 seconds
polonium-216	0.145 seconds

Examples of calculations on half-life

1   What mass of nitrogen-13 would remain if 80 g were allowed to decay for 30 minutes? Nitrogen-13 has a half-life of 10 minutes.

   after 1st 10 min (total time 10 min)   80 g would decay to 40 g
   after 2nd 10 min (total time 20 min)   40 g would decay to 20 g
   after 3rd 10 min (total time 30 min)   20 g would decay to 10 g

   Answer: 10 g would remain after 30 minutes.

2   How long would it take for 20 g of cobalt-60 to decay to 5 g? The half-life of cobalt-60 is 5.26 years.

   20 g to 10 g takes 5.26 years
   10 g to 5 g takes another 5.26 years

   Answer: Total time taken is 10.52 years.

3   Strontium-93 takes 32 minutes to decay to 6.25% of its original mass. Calculate the value of its half-life.

   100% to 50% takes 1 half-life
   50% to 25% takes 1 half-life
   25% to 12.5% takes 1 half-life
   12.5% to 6.25% takes 1 half-life
   so 4 half-lives take 32 minutes, each half-life = 32/4 minutes

   Answer: The half-life of strontium-93 is 8 minutes.

Exam questions

1 The carbon-14 isotope is radioactive and undergoes beta decay.

The half-life of the carbon-14 isotope is about 5700 years.

a) Give the atomic number and the mass number of the species formed when carbon-14 undergoes beta decay.
(2 marks)

b) How much of a 20 g sample of carbon-14 would remain after 17 190 years? Show how you arrive at your answer.
(2 marks)

c) The presence of the carbon-14 isotope is used to date old materials such as wooden boats and fabrics such as the 'Turin Shroud'. Explain why the carbon-14 isotope is a very suitable choice for dating such materials.
(4 marks)

2 A radioactive isotope X undergoes beta. decay.

a) Give the charge and the mass of a beta particle.
(2 marks)

b) If 8 g of the isotope X decays to 1 g after 12 days, find the half-life of X. Show how you arrive at your answer.
(2 marks)

c) Explain why isotope X would not be a very suitable radioactive source for dating old objects such as rocks, assuming the rocks originally contained some X.
(2 marks)

3 a) All living things contain the radioactive isotope carbon-14 which undergoes beta decay. Copy and complete the nuclear equation below to show what happens when carbon-14 undergoes beta decay.

$$^{14}_{6}C \rightarrow \qquad + $$
(2 marks)

b) A very old wooden boat was found to contain only 25% of the carbon-14 which is present in a living tree. The half-life of carbon-14 is 5730 years. How long ago was the wood cut down to make the boat? Show your working.
(2 marks)

c) Nuclear power can be generated by the fission of uranium-235. Describe what happens during nuclear fission.
(2 marks)

4 a) Uranium consists of two isotopes, uranium-235 and uranium-238.

(i) What name is given to the process when a neutron strikes a uranium nucleus and splits it into two fragments with the release of a large amount of energy?
(1 mark)

(ii) Copy and complete the nuclear equation to show what happens when uranium-238 loses an alpha particle.

$$^{238}_{92}U \rightarrow {}^{4}_{2}He + $$
(2 marks)

211

b) The table below shows how the percentage of a sample of iodine-131 changes with time.

Time (days)	0	5	10	15	20	25	30
% iodine-131	100	68	40	26	18	11	7

(i) Plot the points on a graph like the one below and draw the curve.

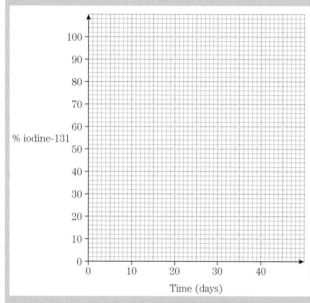

(*3 marks*)

From this graph the half-life of iodine-131 can be found.

(ii) Explain what is meant by the term **half-life**.

(*2 marks*)

(iii) What is the half-life of iodine-131, to the nearest day?

(*1 mark*)

Chapter 18

Calculations

By the end of this chapter you will:

➤ Have learnt about the relative atomic mass and how to use it to work out relative formula mass

➤ Understand the concept of the mole, which is a word for the amount of a chemical

➤ Know how to use the mole to find out the mass of a product formed or of a reactant used in a reaction

➤ Know how to work out the formula of a compound

➤ Know what is meant by the concentration of solutions

➤ Know how to find the concentration of a solution by an experiment followed by a calculation

➤ Know how to work out the volume of gases involved in reactions

Relative atomic mass (RAM)

Atoms are extremely small and so are their masses. As a result chemists do not use actual masses. The mass of the carbon-12 isotope can be measured very accurately and the masses all other atoms are compared with this to give a scale of relative atomic masses.

The carbon-12 isotope is given a RAM of 12.
The hydrogen atom weighs $\frac{1}{12}$ of carbon-12 and so it has a RAM of $\frac{12}{12} = 1$
The magnesium atom weighs twice as much as a carbon-12 and so it has a RAM of $12 \times 2 = 24$

$$\text{Relative atomic mass} = \frac{\text{actual mass of atom}}{\frac{1}{12} \text{ of carbon-12}}$$

You can find the values of the relative atomic masses on your Periodic Table, it is the top number in each symbol.

Relative formula mass, RFM (relative molecular mass, RMM)

(There are two names because not all elements or compounds exist as molecules but they all have formulae.)

The **relative formula mass** can be found by adding together all the atomic masses of all the atoms in the formula of the element or compound.

Examples

Br_2:	$RFM = (RAM\ Br) \times 2 = 160$
NaCl:	$RFM = RAM\ Na + RAM\ Cl = 58.5$
Na_2CO_3:	$RFM = (RAM\ Na \times 2) + RAM\ C + (RAM\ O \times 3) = 106$
$Ba(NO_3)_2$:	$RFM = RAM\ Ba + [RAM\ N + (RAM\ O \times 3)] \times 2 = 261$
$CuSO_4.5H_2O$:	$RFM = RAM\ Cu + RAM\ S + (RAM\ O \times 4)$
	$\quad\quad + 5[(RAM\ H \times 2) + (RAM\ O)] = 250$

Questions

1 Calculate the relative formula mass (relative molecular mass) of each of the following:

a) NaOH b) $CaCO_3$ c) $Al(OH)_3$

d) $(NH_4)_2SO_4$ e) $CoCl_2.2H_2O$

The mole

The mole is an amount of a chemical.

Moles and numbers of particles

One mole of any element or compound contains the same number of particles, which is 6×10^{23}. The number of moles is found by dividing the number of particles (atoms or molecules) by 6×10^{23}.

$$\text{moles} = \frac{\text{number of particles}}{6 \times 10^{23}}$$

Examples

1 How many moles are in 3×10^{23} atoms of tin?

$$\text{moles} = \frac{3 \times 10^{23}}{6 \times 10^{23}} = 0.5$$

2 How many molecules are in 2 moles of water?

$$\text{no. of molecules} = 2 \times 6 \times 10^{23} = 1.2 \times 10^{24} \text{ or } (12 \times 10^{24})$$

This very large number is known as the **Avogadro constant** and is often represented by the letter L. The type of particle depends on the element or compound concerned.

● Potassium metal exists as atoms so 1 mole K contains L atoms.

● Chlorine gas exists as molecules so 1 mole of Cl_2 contains L molecules.

● Water exists as molecules so 3 moles of H_2O contains 3L molecules.

● Argon exists as atoms so $\frac{1}{2}$ mole of Ar contains $\frac{1}{2}$ L atoms.

Moles and mass

One mole of any element or compound weighs its RAM or RFM in grams and so the number of moles in a sample can be found

$$\text{moles} = \frac{\text{mass (g)}}{\text{RAM or RFM}}$$

If the moles are known then the mass can be found

$$\text{mass} = \text{moles} \times \text{RAM/RFM}$$

Examples

1   How many moles in 4 g sodium hydroxide?

$$\text{moles NaOH} = \text{mass/RFM} = \frac{4}{40} = 0.1$$

2   What is the mass of 3 moles ammonia?

$$\text{moles NH}_3 = \text{moles} \times \text{RFM} = 3 \times 17 = 51 \text{ g}$$

Questions

2   How many moles are in each of the following?
   a) 2 g of calcium metal   b) 73 g of hydrochloric acid

3   What is the mass of each of the following?
   a) 0.5 moles of nitric acid   b) 0.25 moles of copper oxide

4   a) How many molecules are present in 0.9 g of water?
   b) What is the mass of 3×10^{23} molecules of carbon dioxide?

Using moles and masses

Calculating reacting masses

Mole calculations can be used to find the masses of reactants or products involved in reactions. These calculations use a balanced equation to find the mole ratio of the chemicals involved and then moles = mass/RFM and mass = moles $\times$ RFM.

Examples

1   What mass of copper oxide would be obtained from the thermal decomposition of 6.2 g of copper carbonate?

$$\text{CuCO}_3 \rightarrow \text{CuO} + \text{CO}_2$$
$$1 \text{ mole} \qquad 1 \text{ mole}$$

$$\begin{array}{lll}
\text{moles CuCO}_3 = \text{mass/RFM} & = 6.2/124 & = 0.05 \\
\text{moles CuO} & = \text{moles CuCO}_3 = 0.05 \\
\text{mass CuO} & = \text{moles} \times \text{RFM} = 0.05 \times 80 = 4
\end{array}$$

2 What mass of potassium hydroxide would be required to react with excess sulphuric acid in order to produce 1.74 g of potassium sulphate?

$$2KOH \; + \; H_2SO_4 \; \rightarrow \; K_2SO_4 \; + \; 2H_2O$$
$$\text{2 moles} \qquad \text{1 mole}$$

moles K_2SO_4 = mass/RFM   = 1.74/174 = 0.01
moles KOH   = moles $K_2SO_4 \times 2$ = 0.01×2 = 0.02
mass KOH   = moles $\times$ RFM   = 0.02×56 = 1.12 g

Questions

5 Calculate the maximum mass of carbon dioxide that could be obtained from 6.4 g of potassium carbonate reacting with excess hydrochloric acid.

$$K_2CO_3 + 2HCl \rightarrow 2KCl + CO_2 + H_2O$$

6 What would be the minimum mass of calcium oxide that would be needed to react with excess water to give 9.25 g of calcium hydroxide?

$$CaO + H_2O \rightarrow Ca(OH)_2$$

Calculating the formula of a simple compound

The simplest formula for a compound is known as the **empirical formula**. It gives the elements present and their ratio.

Examples

NaCl shows that sodium and chlorine are present in the ratio of 1:1
H_2O shows that hydrogen and oxygen are present in the ratio 2:1

As equal numbers of moles of elements contain equal numbers of atoms then the ratio of moles can be used to find the ratio of atoms.

1 molecule H_2O = 2 atoms H   + 1 atom O
10 molecules H_2O = 20 atoms H   + 10 atoms O
6×10^{23} molecules H_2O = 1.2×10^{24} atoms H + 6×10^{23} atoms O
1 mole H_2O = 2 moles H   + 1 mole O

The ratio of atoms in a molecule is the same as the ratio of moles of the elements in a sample of the substance.

Examples

1 A chloride of magnesium contained 0.24 g of magnesium and 0.71 g of chlorine. Calculate its empirical formula.

As the mass of each element is given the moles can be found using the equation moles = mass/RAM

	Mg	Cl
Mass (g)	0.24	0.71
RAM	24	35.5
Moles	$\dfrac{0.24}{24} = 0.01$	$\dfrac{0.71}{35.5} = 0.02$

The next stage is to find the ratio of the moles, look for the smallest number of moles and divide all the moles by that number. In this example the magnesium has the smaller number of moles

	Mg	Cl
Ratio	$\dfrac{0.01}{0.01} = 1$	$\dfrac{0.02}{0.01} = 2$

The ratio of moles is the same as the ratio of the numbers of atoms provided this gives whole numbers of atoms. Remember that you cannot have half, or any other fraction, of an atom.

	Mg	Cl
Atoms	1	2 (N.B. whole numbers)

Formula = $MgCl_2$

2 1.60 g of an oxide of iron was found to contain 1.12 g of iron. Calculate the empirical formula of the oxide.

The compound contains iron and oxygen only so the mass of oxygen can be found by subtracting the mass of iron from the mass of the iron oxide.

	Fe	O
Mass (g)	1.12	$1.60 - 1.12 = 0.48$
RAM	56	16
Moles	$\dfrac{1.12}{56} = 0.02$	$\dfrac{0.48}{16} = 0.03$
Ratio	$\dfrac{0.02}{0.02} = 1$	$\dfrac{0.03}{0.02} = 1.5$

You cannot have 0.5 of an atom in a compound so double the numbers to get 2:3 instead of 1:1.5

	Fe	O
Atoms	2	3 (N.B. whole numbers)

Formula = Fe_2O_3

3 A sample of a compound was found to consist of 7.2 g of carbon and 1.6 g of hydrogen only. Calculate its empirical formula.

	C	H
Mass (g)	7.2	1.6
RAM	12	1
Moles	$\dfrac{7.2}{12} = 0.6$	$\dfrac{1.6}{1} = 1.6$
Ratio	$\dfrac{0.6}{0.6} = 1$	$\dfrac{1.6}{0.6} = 2.67$

You cannot have 0.67 of an atom. Doubling gives 2:5.34, still no good. Multiplying by 3 gives 3:8.01, this is close enough to 3:8

	C	H
Atoms	3	8 (N.B. whole numbers)

Formula = C_3H_8

7 A compound contains 1.03 g of lead and 0.80 g of bromine. What is its empirical formula?

8 A sample of a compound of carbon and hydrogen only weighs 1.22 g and contains 1.20 g of carbon. What is its empirical formula?

Moles and concentration

The concentration of a solution is a measure of the amount of solute dissolved in a certain volume of solution.

The usual amount is measured in moles (mol) and the usual volume is measured in decimetres cubed (dm^3) or litre (l). The units of concentration are therefore: **mol/dm³** or **mol/l**.

Calculating concentration

$$\text{moles} = \text{conc. (mol/dm}^3) \times \text{vol (dm}^3)$$

Examples

1 How many moles of sulphuric acid are in 5 dm^3 of a 0.1 mol/dm³ solution?

$$\text{moles} = 0.1 \times 5 = 0.5 \text{ mol}$$

2 What will be the concentration of a 2 dm^3 solution which contains 4 moles of potassium hydroxide?

$$\text{concentration} = 4/2 = 2 \text{ mol/dm}^3$$

To make up a solution of known concentration, **a standard solution**, it is necessary to work out the mass of solute that has to be weighed to give the number of moles needed.

Example

What mass of anhydrous sodium carbonate is required to give 250 cm^3 of a 0.05 mol/dm³ solution?

$$\text{vol in dm}^3 = \frac{250}{1000} = 0.25 \text{ dm}^3$$

$$\text{mol Na}_2\text{CO}_3 = \text{conc. (mol/dm}^3) \times \text{vol (dm}^3)$$

$$= 0.05 \times 0.25 = 0.0125 \text{ mol.}$$

$$\text{Mass Na}_2\text{CO}_3 = \text{mol} \times \text{RFM} = 0.0125 \times 106$$

$$= 1.325 \text{ g}$$

Questions

9 What volume of 0.25 mol/dm^3 potassium hydroxide solution is needed to provide 0.5 moles of potassium hydroxide?

10 Calculate the mass of sodium nitrate needed to make 250 cm^3 of 0.1 mol/dm^3 solution.

Titrations

An acid–alkali titration involves reacting a solution of an acid with a solution of an alkali, in a conical flask, until the mixture is neutral. An indicator is used to show when the neutral point, known as the end point, has been reached.

After carrying out a titration the volumes of both solutions are known and the concentration of one, the standard solution, is also known. The concentration of the second solution can be found by calculation.

The calculation can be carried out using a formula for titration calculations based on moles = volume × concentration and the ratio of moles from a balanced equation.

$$aA + bB \rightarrow cC + dD$$

(A, B, C and D are the substances and a, b, c and d are the numbers of moles of each substance.

$$\frac{\text{vol A} \times \text{conc A}}{\text{vol B} \times \text{conc B}} = \frac{\text{moles A}}{\text{moles B}} = \frac{a}{b}$$

Examples

$$H_2SO_4 + 2KOH \rightarrow K_2SO_4 + 2H_2O$$

$$\frac{\text{vol } H_2SO_4 \times \text{conc } H_2SO_4}{\text{vol KOH} \times \text{conc KOH}} = \frac{\text{moles } H_2SO_4}{\text{moles KOH}} = \frac{1}{2}$$

Unlike using moles = volume × concentration on its own it is not necessary to have the volume in decimetres cubed (or litres) as long as both volumes are in the same units.

Examples

1 A 25.0 cm^3 sample of a 0.1 mol/dm^3 solution of sodium hydroxide required 24.6 cm^3 of a sulphuric acid solution for complete neutralisation. Calculate the concentration of the sulphuric acid.

$$H_2SO_4 + 2NaOH \rightarrow Na_2SO_4 + 2H_2O$$
1 mole   2 moles

$$\frac{\text{vol } H_2SO_4 \times \text{conc } H_2SO_4}{\text{vol NaOH} \times \text{conc NaOH}} = \frac{\text{moles } H_2SO_4}{\text{moles NaOH}} = \frac{1}{2}$$

$$\frac{24.6 \times \text{conc } H_2SO_4}{25.0 \times 0.1} = \frac{1}{2}$$

$$\text{conc } H_2SO_4 = \frac{1 \times 25.0 \times 0.1}{24.6 \times 2} = 0.05 \text{ mol/dm}^3$$

219

2   A 20.0 cm³ sample of a 0.02 mol/dm³ hydrochloric acid solution required 10.3 cm³ of a potassium hydroxide solution for complete neutralisation. Calculate the concentration of the potassium hydroxide solution.

$$KOH \quad + \quad HCl \qquad \rightarrow KCl + H2O$$
$$\text{1 mole} \qquad\qquad\qquad \text{1 mole}$$

$$\frac{\text{vol KOH} \times \text{conc KOH}}{\text{vol HCl} \times \text{conc HCl}} = \frac{\text{moles KOH}}{\text{moles HCl}} = \frac{1}{1}$$

$$\frac{10.3 \times \text{conc KOH}}{20.0 \times 0.02} = \frac{1}{1}$$

$$\text{conc KOH} = \frac{1 \times 20.0 \times 0.02}{1 \times 10.3} = 0.04 \,\text{mol/dm}^3$$

Questions

11 What is the concentration of a solution of potassium carbonate if 25.0 cm³ of it needs 12.5 cm³ of 0.1 mol/dm³ hydrochloric acid for complete neutralisation?

12 It was found by titration that 10.0 cm³ of 0.1 mol/dm³ sodium hydroxide solution was neutralised by 19.8 cm³ of sulphuric acid. Calculate the concentration of the acid.

Moles and gases

Avogadro's Law

This law states that equal volumes of gases, measured at the same temperature and pressure, contain equal numbers of molecules.

As equal numbers of moles contain equal numbers of molecules Avogadro's Law may be updated to state that equal volumes of gases, measured at the same temperature and pressure, contain equal numbers of moles.

Examples

1   What volume of carbon dioxide would be formed when 12 dm³ of carbon monoxide is completely combusted? All volumes are measured at the same temperature and pressure.

$$2CO + O_2 \rightarrow CO_2$$
$$\text{2 mole} \qquad\quad \text{1 mole}$$
$$\text{2 vol} \qquad\qquad \text{1 vol}$$
$$\text{12 dm}^3 \qquad\quad \text{6 dm}^3$$

2   What is the minimum volume of hydrogen needed, and what is the volume of ammonia formed, when 40 cm³ of nitrogen is reacted with hydrogen? All volumes are measured at the same temperature and pressure.

$$N_2 \quad + \quad 3H_2 \quad \rightarrow \quad 2NH_3$$
$$\text{1 mole} \qquad \text{3 moles} \qquad \text{2 mole}$$
$$\text{1 vol} \qquad\quad \text{3 vol} \qquad\quad \text{2 vol}$$
$$\text{40 cm}^3 \qquad \text{120 cm}^3 \qquad \text{80 cm}^3$$

Questions

13   What is the minimum volume of oxygen that would be needed for the complete combustion of 20 cm^3 of hydrogen? All volumes are measured at the same temperature and pressure.

$$2H_2 + O_2 \rightarrow 2H_2O$$

Molar gas volume

The volume of any gas is mainly the space between the particles. The volume of the particles themselves is a very small fraction of the total volume occupied by the gas and so the volume of any gas depends on the temperature and pressure and not on the nature of the gas.

One mole of any gas, measured at room temperature and pressure, has a volume of 24 dm^3.

$$\text{Moles (of gas at RTP)} = \frac{\text{volume (dm}^3)}{24}$$

$$\text{volume (of gas at RTP)} = \text{moles} \times 24$$

Examples

1   How many moles of ammonia are present in a sample which occupies 6 dm^3 at room temperature and pressure?

$$\text{moles NH}_3 = \frac{6}{24} = 0.25 \text{ mol}$$

2   What volume will be occupied by 0.01 moles of oxygen at room temperature and pressure?

$$\text{vol O}_2 = 0.01 \times 24 = 0.24 \text{ dm}^3 = 240 \text{ cm}^3$$

Questions

14   How many moles are present in 0.3 dm^3 of chlorine gas, measured at room temperature and pressure?

15   Calculate the volume that 0.25 moles of nitrogen will occupy at room temperature and pressure.

16   **IT:** Look at the website www.rod.beavon.clara.net/avo1.htm

Websites

www.s-cool.co.uk

www.rod.beavon.clara.net/avo1.htm

1   A chemist heated 5.08 g of iodine with 1.6 g of tin (Sn) until all the iodine had reacted. Tin iodide was formed and 0.41 g of tin remained unreacted at the end of the experiment. (Relative Atomic Masses: Sn = 119, I = 127)

a) What mass of tin actually reacted with the iodine?   (*1 mark*)

b) How many moles of tin reacted with the iodine?   (*1 mark*)

c) How many moles of iodine atoms reacted?   (*1 mark*)

d) What is the formula of the tin iodide? Show how you work out your answer.   (*2 marks*)

2   To obtain full marks, the steps in the calculations in this question must be shown.

a) Calcium nitrate decomposes on heating according to the equation

$$2Ca(NO_3)_2 \rightarrow 2CaO + 4NO_2 + O_2$$

(i) Calculate the mass of calcium nitrate required to be heated in order to produce 2.8 g of calcium oxide.

(Relative atomic masses: Ca = 40, N = 14, O = 16)   (*4 marks*)

(ii) Calculate the volume of oxygen produced in the same reaction, measured at room temperature and pressure. (1 mole of gas at room temperature and pressure occupies a volume of 24 dm³(l).)   (*3 marks*)

b) (i) State Avogadro's Law   (*3 marks*)

(ii) The hydrocarbon ethane, C_2H_6, undergoes complete combustion according to the following equation.

$$2C_2H_6 + 7O_2 \rightarrow 4CO_2 + 6H_2O$$

What volume of oxygen is required to completely combust 5 dm³ of ethane?   (*2 marks*)

c) A solution of a metal hydroxide, MOH(aq), and dilute hydrochloric acid, HCl(aq), react according to the following equation.

$$MOH + HCl \rightarrow MCl + H_2O$$

The relative formula mass of MOH and hence the relative atomic mass of M can be found by dissolving a known mass of solid MOH in water and titrating it against an acid of known concentration.

In a class experiment, 8.29 g of solid MOH were dissolved in water and made up to 1 dm³ (1000 cm³) of solution. A 25 cm³ sample of this solution required 18.5 cm³ of hydrochloric acid of concentration 0.2 mol/dm³ (moles per litre) for neutralisation.

(i) Calculate the number of moles of hydrochloric acid used in the titration.   (*2 marks*)

(ii) Calculate the number of moles of MOH in the 25 cm³ sample which reacted with the acid.   (*2 marks*)

(iii) How many moles of MOH were in the original 1 dm³ (1000 cm³) of solution?   (*2 marks*)

(iv) Remembering that the original 1 dm³ of solution contained 8.29 g of MOH and using your answer to c)(iii), calculate the relative formula mass of MOH.   (*2 marks*)

(v) Using your answer to c) (iv), calculate the relative atomic mass of M.

(Relative atomic masses: H = 1, O = 16)   (*2 marks*)

(vi) What is element M?   (*1 mark*)

Chapter 19

Chemistry at Work

Learning objectives

In this chapter you will find information about a selection of human activities that affect our world. By the end of it you will:

➤ Know that any one activity has 'good' and 'bad' aspects

➤ Realise that pollution is a widespread problem and that it needs to be dealt with locally and internationally

➤ Know how to evaluate the social, economic and environmental factors that are associated with industrial processes

All human activity has an influence on our environment and especially on our use of the natural resources found on Earth.

Since they first evolved, humans have made use of natural materials. Early humans used plant and animal material for food, coverings and building shelters. They also made use of minerals for tools.

Today we continue to exploit the raw materials found on Earth to satisfy our needs. Some needs are more important than others! They range from basics such as food to luxuries such as diamonds and they include the raw materials required by industry.

The extraction of metals from their ores, the processing of food and the manufacture of pharmaceuticals, dyes, pigments, plastics and fertilisers are only a few examples of 'chemistry at work'.

Some examples of 'chemistry at work'

Limestone quarrying

Limestone, calcium carbonate, has many uses and is a very important raw material for industry.

Limestone is an abundant raw material and as it is relatively easily obtained it is fairly cheap. The business of quarrying provides jobs at the quarry and in transporting the limestone.

Unfortunately, quarries destroy natural habitats. The traffic associated with them contributes to noise and dust pollution, as does the blasting that is part of the process.

If a new quarry is to be dug consideration must be given to the balance between the demands of industry and care for the environment.

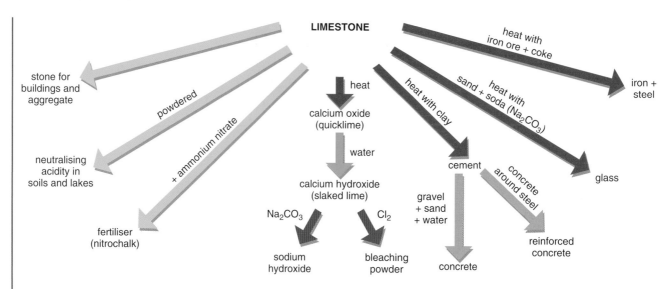

Figure 1 Important uses and products of limestone

Questions

1   Why is limestone considered to be a cheap raw material?

2   Give two advantages and two disadvantages of setting up a quarry.

Peat cutting

Peat has been used as a fuel for centuries. In Ireland families had rights to remove the turf from a certain, usually quite small, strip of the local peat bog. Once industry became involved, however, the removal of peat from the bog speeded up considerably. Not only was it being taken for use as a fuel it was also being used in treating soils and making gardening composts.

Peat bogs are very special habitats and many of the plants found in them are not found anywhere else. Large-scale peat cutting was destroying these habitats too rapidly. Concern about the loss of the peat bogs prompted many gardeners to stop using peat-based products.

Lignite mining

Lignite, a soft brown coal, can be found around Lough Neagh. The Crumlin area was estimated to have reserves of about 200 million tonnes that could be obtained by open cast mining. During the energy crisis of the 1960s interest was shown in these reserves. Three possibly commercial seams were identified and they could have been very important for Northern Ireland. It was suggested that the lignite could be extracted and burned to provide 40 per cent of the electricity needs of the country. This would have been of economic value, as we would have needed to import less oil. The project in opening and running the mine would have provided employment, the forecast was for about 400 new jobs. The lignite found in Northern Ireland is low in sulphur and so it would result in relatively little atmospheric pollution when it was burnt.

There were many factors against the project. Not only would the mining destroy the habitat of local wildlife, it could also be unsightly. The area had a thriving fishing industry and productive agricultural land. It included five listed buildings and there was thought to be a possible problem with underground water supplies. In the end the project was shelved (the disadvantages outweighed the possible advantages), but who knows what may happen in the future.

Questions

3 Imagine you want to open a lignite mine near Crumlin. Write a letter to the Department of the Environment trying to persuade them to give you permission to open the mine.

4 Imagine you are part of a group that is against lignite mining in the area near Lough Neagh. Design a leaflet that could be distributed to local people to encourage them to object to the mine.

Solution mining

Common salt, sodium chloride, is a very important raw material for the chemical industry. Most of our salt comes from Cheshire in England and is brought to the surface by solution mining. Water is pumped down into the rock and the salt dissolves leaving behind the rock. The salt solution, brine, is pumped back to the surface. This method has a much less detrimental effect on the environment than other forms of mining. Unfortunately solution mining weakens the rock and can result in subsidence.

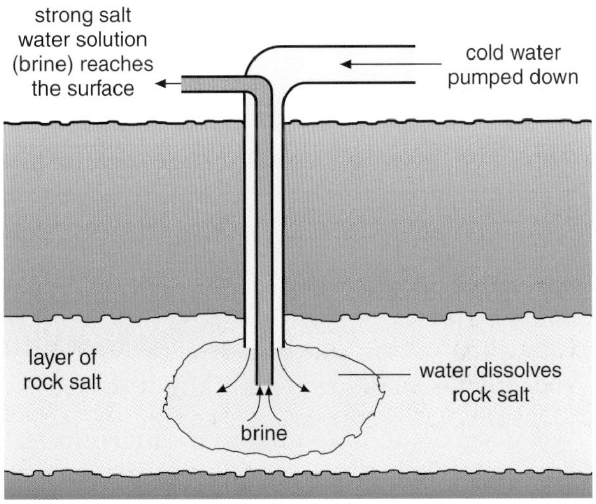

Figure 2 Solution mining

Problems arising from 'chemistry at work'

Pollution

The effect of human activity can result in chemicals being found where they cause harm; such chemical are called pollutants. In many places on Earth our air and water are polluted.

As more has been learned about pollution, efforts have been made to deal with the problems. Once sewage went straight into the local river or the sea but now it is dealt with in sewage works. The Clean Air Act, that allowed local authorities to set up smokeless zones, has greatly reduced air pollution.

Pollution is not, however, a local matter. Oceans, seas and many rivers do not touch only one country. A factory in Basle in Switzerland had a fire, when it was being put out pollutants were washed into the river Rhine. The pollution was carried down the river causing problems as it passed through France, Germany, and The Netherlands (Holland).

The air is definitely international and acid rain is an international problem. The sulphur dioxide may be produced in one country but it is carried away by the wind. Lakes in Scandinavia have been badly affected by acid rain formed in Britain. Greenhouse gases are another matter for international concern.

Governments meet at summits to discuss the environment and the harm we are causing it. Unfortunately getting them all to agree about what has to be done to put matters right seems to be impossible!

Waste disposal

We are creating mountains of waste. Think of all the packaging we use today; plastic waste is a particular problem as very little of it is biodegradable. There are two main ways of dealing with waste; we can either dump it in landfill sites or we can burn it.

Landfill takes up space, can be unsightly and run off can affect local water supplies. Although some of the material dumped in the landfill may decay, much of it will be around for a long time. As a result, the site will become full and a new one will be needed. Finding a new site can be difficult. There can be few people who would want to have one set up near where they live! Old quarries have been used or pits have been dug.

Incineration can be used but not all waste can be burnt. The heat produced could possibly be used to heat water or to provide heating for local buildings. Unfortunately some plastics give off poisonous gases when they are burnt.

Recycling is becoming more and more important, not only to reduce the amount of waste but also to allow non-renewable resources to be re-used.

Websites

www.gcsechemistry.com/ukop.htm

www.soton.ac.uk/~engenvir/environment/air/air.pollution.html-3k

www.bbc.co.uk/weather/features/air_pollution.shtml

Exam questions

1 In the chloralkali industry electricity is passed through a solution of common salt in water to form the products shown in the diagram.

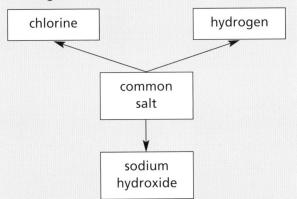

a) Give a balanced, symbol equation for the overall reaction occurring in the process.

(2 marks)

b) From an economic point of view, why is this process so important?

(2 marks)

c) What would you consider to be one of the major costs in the chloralkali industry?

(1 mark)

d) Name the hazard symbol which would be attached to containers of each of the products.

(3 marks)

e) Chlorine reacts with cold, dilute sodium hydroxide solution.
Give a balanced, symbol equation for this reaction.

(2 marks)

2 Peat, or turf, is found throughout Ireland. It is cut into pieces and dried before being used.

a) What is peat used for?

(1 mark)

b) Give **two** advantages of peat cutting.

(2 marks)

c) Give **one** disadvantage of peat cutting.

(1 mark)

3 Sodium chloride and calcium carbonate are extracted from the ground and have important uses. The method of extraction of sodium chloride is very different from that used to extract calcium carbonate.

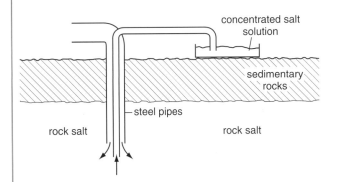

a) (i) Using the diagram above to help you, describe how solid sodium chloride is obtained from the ground by solution mining.

(3 marks)

(ii) Give **one** use for sodium chloride.

(1 mark)

b) (i) Give **two** uses for calcium carbonate.

(2 marks)

(ii) Calcium carbonate is obtained by more conventional mining techniques. Explain why it is **not** obtained by solution mining.

(1 mark)

c) Conventional mining and solution mining can both give rise to environmental problems.

(i) State **two** negative effects of the conventional mining of calcium carbonate.

(2 marks)

(ii) Suggest **one** negative effect connected with the solution mining of sodium chloride.

(1 mark)

Index

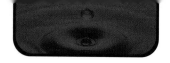